EINSTEIN DIDN'T

SAY THAT

EINSTEIN DIDN'T

SAY THAT

EXPOSING THE COMMON SENSE
IN RELATIVITY THEORY

DON GRIFFIN

ISBN: 978-0-9865552-0-6

Cover Photo by Don Griffin: mirror and salt

ACKNOWLEDGEMENTS:

Thanks to Chris Bruner for suggestions and encouragement, and for so much sharing of ideas over the years. Thanks to my wife Olga for her proofing skills and her judgment, and also for explaining to the computer what I wanted it to do.

Thanks to the Guelph Public Library and its staff for their extensive collection, and for their help with inter-library loan requests. Thanks especially to the Perimeter Institute for Theoretical Physics for keeping me enthused, and to my sister Carole for thinking I could write.

Thanks to all the smart people quoted here, who don't know me, with apologies if I have quoted them too far out of context. Apologies also for any errors, all of which are mine.

TABLE OF CONTENTS

INTRODUCTION

If someone were to tell us that Julius Caesar is just as alive as the rest of us, should we believe it? What if the person who told us this had both a PhD and a Nobel Prize? Well mathematical physicist Sir Roger Penrose does have both, and if I heard correctly, he did say this very thing on a recent radio program.[1] What I think Sir Roger said was that if Einstein's original two assumptions were correct, regarding the behaviour of light and the laws of nature, then using mathematics based on these assumptions, you could come to the conclusion that old Julie is fine and well thanks very much.

How does Sir Roger know this? Well, since it's in the math, it can't be explained to just anybody. It reminds me of the slogan on the back of some Jeeps that reads: "It's a Jeep thing; you wouldn't understand." Well I'm not one who can argue about math or Jeeps. But I would like to be able to tell the difference, all by myself, between what Einstein said, and the ideas he inspired in so many other people.

Yogi Berra purportedly said "I never said half the things I said". This surely applies to Einstein as well. Now if only everyone would be as careful as Dr. Penrose was, in specifying that his conclusions were his own, we would not have to sort out all the

---

1 CBC Radio: "Ideas" hosted by Paul Kennedy.

statements that begin: "According to Einstein….". You see the difference. But thankfully, Einstein did write enough material that was intended for the general public, in his simple, charming and honest style, that we can check some of this stuff for ourselves.

Did Einstein say that time slows down, or that empty space is curved, or that gravity is not a force? For answers we look to the experts. But sometimes the experts don't all answer these questions in the same way, creating the appearance of disagreement. What I think is really happening is that they are all trying to explain the same things from different perspectives, using language which, in the case of English at least, is not always precise enough to be up to the task. They have to pick and choose their words, adapting their approach to what they think we will best understand. No use trying to give us too much all at once, especially if we don't speak math.

The result is that physics has not been packaged and delivered to the public as efficiently or as consistently as most other things in our consumer society, in these last many years. Not like milk, for example, where you can choose your favourite percentages of fat, protein and lactose, depending on which you are able to digest most enjoyably, and all in the same convenient packaging. It's all regulated, too. If you go in to a grocery store and ask for soy milk they will likely direct you to the dairy isle without blinking. The soy milk is beside the cow's milk and the packaging is similar. We all understand each other. But you won't find the word 'milk' on the package, at least not in my neighbourhood. Regulations forbid it. Somebody might get confused. Somebody trying to learn English, or to understand what cows do, might think that cows turn hay and corn into soy products. With physics, there are no such regulations. Confusion is permitted. You can say we live in the milky way and nobody will stop you or make you prove it's made of milk.

So we hear weird sounding language, because some of the old words have taken on new meaning, and some of the new words haven't been invented yet, and we hear that it is all based upon Einstein's ideas. Wanting to understand Einstein, we ask about the weird sounding stuff. As a result we get weird sounding answers to the wrong questions. The fundamental questions don't get asked or answered directly. We yearn for access to the logical thinking which we know must lie at the bottom of it all, but rather than finding it all in one place, we have to search for it among clues inadvertently buried in stories about more newsworthy stuff.

Sometimes we don't find the clues at all. Or we confuse the clues with the fundamentals. Does curved space cause gravity, or does gravity make space curve? Sometimes the clues are contradictory. Does time really slow down, or not really? Why do we hear it both ways? Or worse yet, we assume that the real logic is all in the math, way beyond our understanding, and we just go back to playing with our crayons.

It seemed odd to me that, along with their descriptions of

things that sounded weird, the experts would often mention, in a matter of fact sort of way, that Einstein was unwilling to accept concepts that went against his common sense. It didn't make sense to Einstein that gravity could act on things at a distance, without a physical link. And it didn't make sense that gravity could act on distant things instantaneously, even faster than the speed of light. And things couldn't depend on chance; God didn't play dice.

There seemed to be a double message coming from the experts. Physics could be weird, even when it was based on Einstein's ideas, but Einstein insisted on things not being weird. For me, things didn't begin to sort themselves out until I read Einstein's own explanation of his fundamental ideas, written in his own straightforward style. In 1916 Einstein wrote a book for a general audience, entitled simply: 'Relativity'. In it he explains the 'empirical physical foundations of the theory'.[2]

Einstein had found a way to make sense of things that were already known about nature, but that seemed contradictory. It was all about making things fit with common sense. Finally I began to realize that the weirdness wasn't in Einstein's fundamental ideas at all; it was more a reflection of the inadequacy of our language. The more we know about things the more words we need to describe them. Apparently the Inuit have a hundred different words for 'snow'. Our language couldn't easily adapt to new ways of looking at things like time and space and gravity.

_____

2 Albert Einstein: Relativity, p v.

Even the word 'relativity' itself does not really fit; Einstein preferred the word 'invariance', since the laws of nature are the same everywhere. If this had caught on, we wouldn't hear the common phrase 'everything is relative', being applied to everything from the price of potatoes to social values and morality. It is unlikely that we would often use the phrase: 'everything is invariant'. We already know that we can't change the laws of nature. Maybe we just don't want to admit it, or maybe it's just more fun to think that we can. Einstein ends the preface to his little book with the words:

> "May the book bring some one a few happy hours
> of suggestive thought."[3]

It feels like he is speaking to me, rather than to the thousands of scientists and millions of others who have studied his work for more than a century. What follows is a description of my own share of those happy hours.

---

3 Albert Einstein: Relativity, p vi.

Einstein Didn't Say That

Chapter 1

CURVED SPACE

Space is curved. Thanks to Einstein everyone knows this. This is Einstein's theory of gravity, right? Objects are attracted to each other because space is curved. At least this is what I had thought people were telling me that Einstein said. Here is part of what one Einstein biographer had to say:

> "Gravity, he said, isn't a force at all; it's a curvature in the fabric of space-time, and objects must follow that curvature."[4]

And now a physicist:

> "According to Einstein, there is no gravitational pull. The earth warps the space-time continuum around our bodies, so space itself pushes us down to the floor."[5]

---

4 Jurgen Neffe: Einstein, A Biography. P41.

5 Michio Kaku: Einstein's Cosmos, p98.

Notice the subtle difference. The physicist does not say that gravity is not a force. The physicist is explaining curvature, but not how the earth creates it. But pushing is involved. And so I am left wanting to see for myself what Einstein said about curvature. Am I the only one who doesn't understand how curved space can make objects fall to earth? Am I the only one who can't feel curved space pushing down on my head? Does curved space squeeze things together? How? How sharply does space have to curve, in order to make a ball which I throw straight up into the air fall straight back down? That's some curve. Curved space. I wanted to believe in it. But first I wanted to see it for myself.

Many of us have seen a demonstration in which a piece of fabric is stretched out and suspended, and a heavy ball is placed in the center. We are asked to imagine that the fabric is space and the heavy ball is a planet or a sun. There is also a smaller ball, usually one of those green tennis balls, sitting on one of the corners of the fabric. Then the lecturer gives the tennis ball a little push, as if to make it roll across one end of the fabric. But instead it rolls toward the middle where the big ball is bulging down like Grandpa in the backyard hammock, and, depending upon the strength of the lecturer's shove, the tennis ball either curves a little toward the heavier ball and goes on by, or it rolls to a stop and snuggles down beside Grandpa. An effective and satisfying demonstration it is. We have watched the tennis ball follow the curvature of space, and we now understand gravity and general relativity. Or do we?

I for one am left staring at the little ball and wondering what would have happened if the lecturer had not given it a shove. Why didn't the ball move without being pushed, like the apple falls from the tree? To take a different example for a moment, consider that a train with a good engine will move forward and follow a track. If the track is curved, the engine will pull the train around the curve. But if the train has no engine there will be no motion at all. The curved track can't pull the train. No pull, no motion. Similarly with the tennis ball; no shove by the lecturer means no motion, and no demonstration. Grandpa is still sleeping soundly in his hammock undisturbed by tennis balls. This flexible sheet thing always left me unsatisfied..

And so it was that I found myself driving to yet another public lecture, hoping that maybe this time I might get it. Does curvature cause motion? Does space push on stuff? How? How exactly, really? As I drove I was listening to words from Einstein's little book being read to me through my mp3 player by various people with lovely voices and intelligent sounding accents. It was the book Einstein had written for the general public. To be more truthful I should say that I was half listening to it, as one sometimes does while driving. The voices faded in and out of my attention; first a man, then a woman, then a different woman. Suddenly a phrase came from the earphones which made me cry out and take my hands off the wheel . Anyone watching might have mistaken my actions for road rage. "What!?" I yelled at the voice. "Excuse me!?" And then "What!?" again, as if expecting the reader to hear me and explain what she had just read, or apologize for something.

The words which had had this effect on me were: "This acceleration, or curvature … " That was it. Just the beginning of a sentence which already contained too much for me to handle. How do you get from acceleration to curvature just like that? "This acceleration, or curvature … " The phrase made it sound like Einstein was saying that "acceleration" and "curvature" were two different words for the same thing. Well if that is the case, then why not just say that space is accelerated? Not quite as catchy a phrase, is it? What kind of sense would it make to say that space is accelerated? I told myself there must be an error. Maybe the reader had made a mistake. But the more I tried to make it go away, the more the phrase rang in my ears. "This acceleration, or curvature … ".

I turned off the player and made it safely to the lecture. Sure enough, out came the big ball and the little ball. There was the inevitable shove. The bulging plastic sheet made the little ball go toward the big ball. We all clapped as usual, and as usual I was starting to get a headache. I didn't get a chance to ask the lecturer whether curvature actually caused motion or just directed it, but I did at least get to vent afterward on my friend Chris, who listened patiently and paid for the beer.

Later I checked my hard copy of Einstein's little book to see in what context the phrase was used. The words I had heard while driving were in the book all right. Einstein had said "This acceleration, or curvature … ". So I looked on the previous pages.

Here Einstein had been describing a thought experiment involving a man in a large box, the size of an elevator, way out in space far away from any gravitation. The box is then pulled up, (we don't know by whom) with a rope hooked to the top of the box, like an elevator cable. The speed of the upward motion increases steadily. The man inside the box doesn't know about the rope so he thinks he is feeling gravity. Really he is feeling acceleration. Since the man cannot tell the difference, (and neither could we if we were in his shoes), then gravity and acceleration are equivalent. So far so good. That much was easy enough for me to understand, even though I probably never would have thought of it on my own. Gravity and acceleration feel the same. I get it. I've been in an elevator going up, and a plane taking off.

So where does curvature come in? Well, that is easy too. If you shine a ray of light across the box in front of the man, then because the box is not only moving upward, but moving upward faster and faster, the light will be lower in the box by the time it hits the wall on the far side. The man sees the light travel in a curved path inside the box. Yes, this would also work with a little green tennis ball. So what is our friend in the box to think when he sees the light curve as it moves through space in front of him? He might very reasonably assume that the space in the box is curved, and that the moving light has just followed the curve.

Now if only we could call the man in the box on his cell phone, we could tell him why the light beam followed a curved path. We could tell him that there is a rope hooked to the top of the box, and somebody up there is pulling faster and faster on it. What he sees as curvature we see as acceleration. After reading about the man in the box, and how the curvature corresponded to the acceleration, I was able to absorb the whole sentence from Einstein's book:

> "This acceleration or curvature corresponds to the influence on the moving body of the gravitational field prevailing relatively to k1."[6] (k1 means the box the man is in.)

Two things which are indistinguishable, and which Einstein named when he stated his 'equivalence principle', are gravity and acceleration. The realization of this equivalence was what Einstein called "the happiest thought of my life"[7]. This happy thought is common sense at its best. If you remove the chair you are sitting on, your acceleration rate toward the earth will depend on the strength of gravity. Conversely, the feeling of heaviness you get

---

6 Relativity , by Albert Einstein,1931 edition.

7 as quoted in 'Subtle Is The Lord' by Abraham Pais, pg 177.

from gravity could be caused by a corresponding amount of acceleration, as with the man in the box. Here's Einstein again:

> "It can easily be shown that the path of the same ray of light is no longer a straight line when we consider it with reference to the accelerated chest (reference body K1). From this we conclude, that, in general, rays of light are propagated curvilinear in gravitational fields." [8]

If acceleration curves light, and if acceleration and gravity are equivalent, then gravity will also curve light, and by a predictable amount. The famous 1919 observation of exactly this phenomenon is what proved that Einstein's calculations were correct. This made Einstein an instant international superstar.

Notice that Einstein didn't say that the acceleration of the box was caused by a curvature of the space in the box. Einstein said that the acceleration of the box was caused by someone pulling on a rope. A force, not a curvature, caused the acceleration. Similarly a force, not a curvature, causes gravity. Whether or not we use the word 'force' for gravity itself is arbitrary, which, in my opinion, demonstrates a weakness in our language.

Using the word 'gravity' to describe an effect rather than a force explains why some people say that gravity is not a force. But

---

8 Albert Einstein, "Relativity", p 84.

then what word will we use for the force itself? We don't have one. We are accustomed to using 'gravity' for both the force and the effect of the force. No wonder there is misunderstanding. We need more words.

Einstein used 'gravitational fields', comparing them to magnetic fields. Since Einstein's common sense wouldn't allow for action at a distance, he assumed that the magnet could not act directly on an object without an intermediary. In the case of the magnet, the intermediary is the magnetic field. In the case of the man in the box, the intermediary is the rope. In the case of gravity, the intermediary is the gravitational field. Referring to the concept of fields as it applied to both magnetism and gravity, Einstein says that the adoption of this concept is 'arbitrary', which I take to mean that it is accepted on the basis of common sense alone:

> "We shall not discuss here the justification for this incidental conception, which indeed is a somewhat arbitrary one."[9]

Einstein said that if light curves in an accelerated field, like the box on a rope, then light will also curve in a gravitational field, since gravity and acceleration are equivalent. The fact that gravity and acceleration are equivalent also enabled Einstein to predict, very precisely, the amount of curvature of light in a gravitational

---

9   Albert Einstein: Relativity, p 71.

field. The effect of the gravitational field on light in space was identical to the effect that a corresponding amount of acceleration would have on the light in the box. But in each case the force originated in another body, not in the curvature of space.

The first person to say that space was curved was not a scientist at all, but a reporter for the New York Times, who needed a catchy headline for the story about the 1919 experiment. What the experiment proved was that a light ray would bend as much as Einstein said it would, rather than as much as Newton said it would. The reporter must not have been aware that Newton's theory could also predict that light would bend, but by a different amount. Within one day after publication of this headline, millions of people had heard the titillating phrase 'space is curved'. The poor physicists got stuck with the job of explaining what this meant, and why Einstein was correct and Newton was wrong. Out came the rubber sheets and the tennis balls. Not that there's anything wrong with that, as the saying goes. It's just that sometimes simple folk like me get the wrong idea, and think that they are saying that curved space causes gravity.

Saying that curved space causes gravity is like saying that ropes cause acceleration. If this were true I would keep a thick rope under the hood of my car. If the man in the box is to feel acceleration, and if we want to know how much the light in the box will curve, we will need not only the rope, but someone to pull on it and lots and lots of math to figure out the amount of curvature. Apparently we would need to use a special kind of geometry that is so difficult that even Einstein had to get help with it.

But this doesn't mean that we need to feel left out. Reading about Einstein's thoughts in Einstein's words, we only need to understand that by leaning back in his chair, he had the happy thought that gravity felt just like acceleration. Then he thought about the man in the box, and how he would see a ray of light curve in front of him.

Did this mean that space itself is curved and is pushing down on us? Maybe, maybe not. But if you ask me, Einstein didn't say that. And if you ask me, we need more words to differentiate between the different elements involved in gravity. Right now we, at least those of us who are not physicists, commonly use the single word 'gravity' to refer to the cause of the force, the force itself, the field the force creates, and the curvature of that field. The rubber sheet demonstration describes only the curvature.

It was a relief to get an answer to the curved space question, even if part of that answer was that nobody really knows exactly how a large body creates a gravitational field, or what the field is composed of physically, or why gravity and acceleration are equivalent. Mainly, it was a joy to realize that common sense answers were available from Einstein himself, along with such honesty and clarity about what was known, what was not known, and what was assumed. I was ready to go back for more, and I was willing to begin at the beginning.

What I had heard about relativity theory, besides that it was weird, which I was beginning to doubt, was that it was based upon two assumptions. The first of these assumptions had to do with the speed of light. And what I had heard about this first assumption was that it had been based upon the results of a famous experiment performed by two scientists named Michelson and Morley.

Einstein Didn't Say That

Chapter 2

MICHELSON-MORLEY EXPERIMENT

I once had a philosophy of science professor who liked to ask his first year classes to find out why the quality of the local water supply didn't change as the level of the underground springs rose and fell with the seasons, or some such question, I forget exactly. The more complicated the students' answers were, the happier this teacher was when he explained that the town water didn't come from underground springs. It was a trick; he wanted to teach us to question the question before looking for an answer. He used this gimmick year after year until word spread and the kids got wise to it. It was a good lesson, even though we may have said otherwise at the time, out of embarrassment. In any case, after that we were a little more careful to look for assumptions buried in questions.

Sometimes people ask why Einstein didn't mention Michelson and Morley in his first paper on Special Relativity. The assumption included in the question is of course that there was a reason that he should have done so. Michelson and Morley had done an experiment involving light. The assumption often made about this experiment is that it proved that the speed of light is constant, even for observers in motion. The further assumption of course is that Einstein knew about it. But now compare this to a quote from Einstein himself, that I found in the classic book "Subtle Is The Lord", written by Einstein's friend and colleague Abraham Pais. Pais writes:

> "In a letter to an historian written a year before his death, Einstein expressed himself for the last time on this subject: 'In my own development, Michelson's result has not had a considerable influence. I do not even remember if I knew of it at all when I wrote my first paper on the subject (1905). The explanation is that I was, for general reasons, firmly convinced that there does not exist absolute motion and my problem was only how this could be reconciled with our knowledge of electrodynamics. One can therefore understand why in my personal struggle Michelson's experiment played no role, or at least no decisive role'.". [10]

But even more pertinent than whether or not Einstein had heard of the experiment, is the question of whether or not it is needed to support Einstein's assumption that the speed of light is constant in all inertial frames of reference, even for observers in motion.

_____

10 "Subtle Is The Lord", by Abraham Pais, p.172.

Michelson and Morley were looking for the ether. The ether idea had long been accepted because people thought that there must be some substance which enabled light to travel, the way the air allows sound to travel. But no one had proved it yet, so this is what they were going to do. Michelson and Morley set out to prove the existence of the ether, using a beam of light. They split a beam of light and then brought it back together, using mirrors. Both halves of the beam traveled the same distance, but one half traveled in a different direction for a bit. The idea was to find out if light would travel at different speeds in different directions due to the ether. When the halves recombined, the original wave pattern was still intact, showing that both halves hit the target at exactly the same instant. If there had been the slightest difference in the speed of the two beams of light, the experimenters would have seen interference in the wave pattern due to a difference in their speed.

Apparently the procedure was quite sophisticated for its day, and even involved stopping traffic nearby to avoid jiggling the equipment. But the idea behind it was simple enough. If the earth is moving through the ether while it spins on its axis, then there should be an 'ether wind' which would produce drag, and affect the speed of light in one direction. Depending on which side of the earth you were on at a given time, the spin of the earth would either add to, or subtract from, the speed of this ether wind as the earth traveled through space, and consequently affect the speed of a beam of light.

Well the result surprised Michelson and Morley, who had expected to find the ether. But it was conclusive; there was no drag on their light beam. They shone it every which way; with the spin of the earth, against the spin of the earth, and across it. They could find no ether wind affecting the speed of light. No ether wind, no ether filling space. And if there was no ether, then it would be reasonable to assume that space was empty. A vacuum. In fact when I was in grade school we were taught that the earth's atmosphere was not very thick, and that beyond the earth's atmosphere was the vacuum of space. But things have changed.

Not too long ago I was at a lecture being given by a young physicist who was describing what was known about dark matter, among other exotic subjects, such as colliding universes. At the end there was a chance to ask questions, so I took the opportunity to ambush her, a little unfairly I admit, and ask her about the speed of light in a vacuum. I say ambush, because the question was a bit off topic, even though she had been talking about there being no vacuum in space. The exchange went something like this:

"So, are you saying that with all this dark matter there really is no such thing as empty space?"

"Yes."

"So you mean you can't have a vacuum?"

"That's right, you can't have a vacuum."

"So if you can't have a vacuum, then how do we know what is the speed of light in a vacuum?"

Well the audience laughed, and the lecturer just lifted her shoulders and looked helplessly at this guy beside her at the front of the room, as if to ask him to step in. I don't know who he was. He said to me: "Come to the next lecture." I sat down, trying not to look as embarrassed as I felt, and wondering why they had not mentioned Michelson and Morley. I wasn't quite sure if the audience had laughed because they thought it was a dumb question or because they thought it was a good question. But I figured that if so much of our new science depended upon special relativity, and if special relativity depended upon the speed of light in a vacuum being a constant, then this question must have an answer. Oh, and the next lecture? Yes I went and, no, it wasn't mentioned. Was it too difficult, too easy, or just too off topic?

Nevertheless, in the days leading up to Einstein's introduction of Special Relativity in 1905, it would have been reasonable for people to think that Michelson and Morley had proven that space was a vacuum, and that light traveled at a constant speed in that vacuum, regardless of the speed of the source of the light. So this could partly explain why people would think that the experiment would have influenced Einstein.

But Einstein went a lot further than just assuming that the speed of light in a vacuum was constant. He assumed it was constant for everybody, even if people are moving away from each other. In Einstein's thought experiment introducing Special Relativity, there was the guy on the train, and somebody on the embankment. They were two observers in motion relative to each other. If they each measured the speed of anything else, like the train, or a tree, in relation to themselves, they would get different answers. But if they measured the speed of light they would both get the same result. Michelson and Morley did not test for this. In the Michelson and Morley experiment there were no observers in motion relative to each other. There were two people in a lab, with a beam of light coming from a source which changed its speed, and a light beam that split, went in different directions, and came back together with both halves still traveling at the same speed. Remarkable, yes, if you were looking for an ether wind. But if you were interested in whether light speed would be measured identically by people in motion relative to each other, you wouldn't be much further ahead, ether or no ether.

So now if my old professor were to ask: "Why didn't Einstein mention Michelson and Morley?", I would question the question. Why should Einstein have mentioned them? They didn't really find a vacuum, and they were not dealing with observers in motion relative to each other. Why should Einstein think that they had proved that the speed of light in a vacuum was constant for observers in motion relative to each other? And I hope the old professor would say back to me: "Good, now explain why Einstein did feel justified in assuming that the speed of light in a vacuum had to be constant."

Chapter 3

SPEED OF LIGHT

When Einstein referred to 'what was known about electrodynamics', he was probably not referring to the Michelson & Morley experiment. He was more likely referring to the work of Faraday and Maxwell. Faraday was the first to introduce the idea of fields. Physicist Roger G. Newton states Faraday's idea:

> "Faraday put forward the idea in 1846 that light was simply a rapid oscillation of the electromagnetic field, …"[11]

But Faraday lacked the mathematical skills to develop the idea fully; this was carried out by Maxwell, beginning from about 1855. Here is physicist Michio Kaku describing Maxwell's work on Faraday's idea:

> "Maxwell quickly realized that this cyclical pattern would create a moving train of electric and magnetic fields, all vibrating in unison, each turning into the other in a never-ending wave.

11 Roger G. Newton: From Clockwork to Crapshoot, p 139.

Then he calculated the speed of this wave. To his astonishment, he found that it was the speed of light. Further, in perhaps the most revolutionary statement of the nineteenth century, he claimed that this was light."[12]

So Maxwell showed that the constancy of the speed of light was a result of how light is created. He didn't measure the speed; he calculated it. If you mix electric and magnetic fields you get an electromagnetic wave. The speed of the wave is constant because of the nature of the ingredients. Physicist Lisa Randall sums up the influence Maxwell's work had on Einstein:

"The notion that light traveled with infinite speed was still accepted by many, including Descartes....Einstein credited Maxwell with the origin of the special theory of relativity: Maxwell's electromagnetic theory gave Einstein the insight about the constant speed of light that instigated his monumental work."[13]

We have all most likely heard the story about Einstein riding on his bicycle at the age of 16, wondering what things would look like if he could ride fast enough to catch up with a light beam.

12 Michio Kaku, Einstein's Cosmos, p28.
13 Lisa Randall: Warped Passages, p 155.

So we know that the nature of light had been on his mind for several years before 1905. Maxwell's theory showed Einstein that the speed of light was in a different category than the speed of his bicycle. You can park a bicycle. You can speed it up and slow it down. Also, bicycles don't move by themselves, or travel at a constant speed, just by virtue of the fact that they are bicycles. Even Einstein had to pedal.

Speed is distance per unit of time. Miles per hour, kilometers per second. Count the miles and count the hours and do the arithmetic. Miles divided by hours. Fifty miles divided by one hour gives you fifty miles per hour. When we drive our cars, we can change the miles and the hours to whatever numbers we want, depending on traffic, and the make of the car, and how long we feel like driving. This is what we know about cars. Comedian Stephen Wright said that when a police officer stopped him and asked him if he knew he had been going eighty miles an hour, he said to the officer: "I wasn't going to be out that long".

But what we know about electrodynamics is that light waves are not like cars, or bicycles. With light we still measure distance and time. To get the distance, we count how many waves there are, and check how far apart they are. But when we count the waves and measure the distance between them, we see that the fewer waves there are, the greater the distance between them, and vice-versa. So for any given amount of time, the distance doesn't change. We know that light of different colors has different frequencies; different numbers of waves per second. Lots of waves close together, or fewer waves farther apart. But in total we get the same distance covered in the same time, which means that the speed is constant. Theoretically, that is.

It's like if I pay a little extra on my mortgage each month, I can reduce the number of payments I will have. Theoretically, that is, until my car decides it would be nice if I bought it a new transmission instead of making the extra mortgage payments. There is the theoretical pay down speed, and the actual pay down speed.

The light speed calculated by Maxwell and used by Einstein is a theoretical light speed. That is why Einstein specifies that he is talking about the speed of light in a vacuum. Whether or not we can ever have a vacuum is not important. We don't need the vacuum, to know what the speed of light would be in that vacuum. The light we see through our bedroom window is not traveling in a vacuum, and so its speed may be a bit off.

Light does travel slower in things like water and glass. Light's speed is also affected by gravity. Since gravity is stronger closer to massive objects, light moves more slowly the closer it passes by them. Here's how Einstein explained it:

> "In the second place our result shows that, according to the general theory of relativity, the law of the constancy of the velocity of light in

vacuo, which constitutes one of the two fundamental assumptions in the special theory of relativity and to which we have already frequently referred, cannot claim any unlimited validity. A curvature of rays of light can only take place when the velocity of propagation of light varies with position."[14]

So, while the velocity of light 'in vacuo' is constant, the fact that light can curve, as it did for the man in the box, means that its velocity can change. When anything goes around a curve, one side of it is moving faster than the other side. The song says that the wheels on the bus go 'round and 'round, but as the bus turns left, the wheels on the right side of the bus are going around faster. In the case of light, the gravity from a massive object slows down light rays according to how close the rays are to the object. More drag on one side causes the light to curve, just as a skater can turn by dragging one foot on the ice.

So we have light being propagated at a constant speed due to the nature of the electromagnetic fields from which it forms, and we have things which can subsequently influence that speed. So far we have come across nothing weird or counter-intuitive that I can see. The speed of light in a vacuum is a theoretical, inherent characteristic of light.

---

14 Albert Einstein, Relativity p 85.

This was one of only two assumptions that Einstein made. The other had to do with the 'principle of relativity' as put forward by Galileo. Here are both of the assumptions stated very succinctly for us by physicist Lisa Randall:

"Einstein's postulates then state that:
The laws of physics are the same in all inertial frames.
The speed of light, c, is the same in any inertial frame."[15]

We often hear that special relativity is mysterious or magical. Since we haven't found these characteristics in the speed of light, maybe we should have a look at the 'inertial frames', and see if there is anything spooky or magical hiding inside them.

---

15 Lisa Randall: Warped Passages p 91.

Chapter 4

INERTIAL FRAMES

We may pick up a book on special relativity, and before we get comfortable in our chair, we find ourselves reading about people with names like 'A' and 'B', flying madly off in all directions, like Don Quixote chasing windmills. Sometimes 'A' stays home, while 'B' goes to the moon. Then maybe 'B' comes back before he left, or younger than he was when he left, and maybe shorter, too, or all of the above. It is magic. It is fascinating. And the book we're reading explains that it all has to do with inertial frames. It's like we all leave high school and go in our different directions, and come back to the reunion years later, all looking younger and thinner than each other. Or so I have heard, as I've never actually had to go to one of those things. In any case we rightly attribute these conflicting perceptions to the fact that we are seeing things from different frames of reference. It is no surprise that things look different to us from different frames of reference; we are used to this. We know that things look smaller when they are far away. If two people are far away from each other they look small to each other, and it makes no sense to ask which one is 'really' smaller. We can use Einstein's own description of an inertial body to get an idea of what is meant by the term 'inertial frame':

"As is well known, the fundamental law of the mechanics of Galilei-Newton, which is known as the law of inertia, can be stated thus: A body removed sufficiently far from other bodies continues in a state of rest or of uniform motion in a straight line."[16]

The man in the box with the hook attached in chapter one was in a frame that was not an inertial frame. It was accelerating, being pulled faster and faster by the rope hooked to the top of it. Similarly, the children on the school bus which was turning left were in a frame that was not inertial. They were feeling centrifugal force pushing them against the windows on the right hand side of the bus. Physicists like to call centrifugal force 'acceleration', mainly I guess because it feels the same and the same math applies to both. Moving frames, and frames at rest, are both inertial frames if they don't feel any type acceleration, or deceleration, for that matter.

So how did Einstein know that he could safely assume that the laws of nature and the speed of light would be the same in these frames? He referred to Galileo's 'principle of relativity', which was a simple statement of common sense. Everyday life tells us that the laws of nature are the same in all inertial frames, whether at rest or in motion.

---

16 Albert Einstein: Relativity, p 13.

Here is Einstein's wording, using 'K' and 'K1' to refer to inertial frames:

> "If, relative to K, K1 is a uniformly moving co-ordinate system devoid of rotation, then natural phenomena run their course with respect to K1 according to exactly the same general laws as with respect to K. This statement is called the principle of relativity (in the restricted sense)."[17]

Whether we pour a cup of tea in our living room, or on a steadily moving train, the tea being poured will not know the difference, and will end up in the cup just the same. Light will travel the length of a room in the same amount of time, whether the room itself is in motion or not. These things we know by experience. This is not mysterious. The laws of nature make the tea go into the cup just the same in our living room or on the train. And the laws of nature, as Maxwell proved, also determine the speed of light, regardless of what frame we choose. In the light of these examples, Lisa Randall's wording of Einstein's two assumptions sounds even simpler the second time:

> "The laws of physics are the same in all inertial frames.

---

17 Albert Einstein: Relativity, p 16.

The speed of light, c, is the same in any inertial frame."

So far no goblins. Easy stuff. An inertial frame is a self-contained neighborhood minding its own business, free of pressure or influence from anything outside itself. Whether we say a particular frame is in motion or at rest, the frame is still inertial if it just keeps on doing whatever it is doing.

So far we have not seen the magical effects of special relativity that are so often alluded to, including time slowing down and twins aging differently. Everything looks normal to us in our own neighborhood. Now let's see if we can conjure up magic by setting the neighborhoods in motion.

Chapter 5

MAGIC TIME

American actor Jack Lemmon, when getting ready to perform, liked to say " It's magic time." Time for the audience to look at him and see someone else, to see something happen that wasn't really happening, or to see familiar things from a new perspective. A different frame of reference. Do clocks magically slow down when they travel at high speed, or does it just look that way from our frame of reference?

We have already seen that light, and everything else for that matter, can be slowed down by gravity. We may not know how gravity works, but we do know what it does, and are not surprised to see it drag things down. It is when we hear that clocks slow down in fast moving inertial frames, which are outside the influence of gravity, that we know that the magic show has started.

One popular illustration of this magic effect involves a spaceship which has a beam of light inside it, shining down from the ceiling. There are mirrors on the floor and the ceiling of the spaceship, and the light bounces back and forth between the mirrors at a steady pace. This up and down reflection of the light beam is in effect a clock. Tick tock. To the people inside the spaceship everything looks normal whether they are moving or not. The light reflects from mirror to mirror and back again. Bounce, bounce, tick, tock. But anyone back on the ground can see that as the spaceship moves away, the light goes not only up and down between the mirrors, but also forward with the ship. They count the same number of ticks and tocks as their flying friends, but see the light going farther for each tick and each tock. They also have to wait a little longer for the ticks and tocks, because of the extra time it takes the light to get from the clock to their eyes. No extra ticks or tocks are counted. Same number of ticks and tocks, but more distance covered and more time taken to count the ticks and tocks. So the people on the ground think that their flying friends' clock has slowed down. Tick...tock...If each tick and tock represents one second, then, to the folks at home, the seconds on the spaceship clock seem to be ticking slowly. Their seconds are longer. Magic time. Did the clock on the spaceship 'really' slow down, or just 'appear' to slow down? The funny thing is that people can agree on what is happening, but answer this question differently. Here is what theoretical physicist David Mermin had to say about it:

> "There is by no means unanimity among practicing physicists on this question, and one frequently finds assertions that, for example, moving clocks *appear* to run slowly when measured by stationary ones, or that moving sticks

appear to shrink. But such caution is uncalled for. Moving clocks really do run slowly and moving sticks really do shrink."[18]

Pardon? Come on now, you practicing physicists. Is it unreasonable for us folks back home to expect unanimity on such a fundamental question? Are the effects of motion on clocks and measuring sticks a matter of opinion? We're not choosing colors for the bathroom here. Please don't leave us on our own to find out what is 'really' happening.

Once when Einstein was asked whether clocks 'really' slowed down in a situation like one with the spaceship, he answered that each person's clock 'really' slowed down for the other person. This answer politely reinforces the fact that all motion is relative, just like all distance is relative. We don't ask who is the far one. And we can't ask who is the moving one either, unless we can prove that one or the other is not moving. If we ask if the clock in motion 'really' slows down, our question includes the assumption that the other clock is not in motion.

Isaac Newton, who is known for assuming that space itself is stationary, nevertheless said the following, about the difficulty of establishing whether something in space is moving or not:

---

18 N. David Mermin: It's About Time, p 189-190.

"It may be that there is no body really at rest, to which the places and motions of others may be referred."[19]

There is no contradiction between the idea that space itself may be stationary, and the fact that it is impossible to tell whether or not something is at rest. Until someone finds a spot in the universe that is definitely home plate, and drives a big nail into it, we can't say that anything is at rest. The person whom Einstein said influenced him most on the subject of rest and motion was Ernst Mach. (Mach is the person from whom we get the expressions 'Mach one, Mach two' etc, referring to speeds of aircraft.) Mach expressed his agreement with the ideas of Galileo and Newton very bluntly:

"For me, only relative motion exists".[20]

So this concept was adopted by Einstein, not invented by him. Here's what the eminent physicist Sir Hermann Bondi said about the source of some confusion concerning relative motion:

"When I talk on special relativity, I always say that Einstein's contribution has a name for being

---

19 James Gleick: Isaac Newton, p 194.

20 As quoted by Abraham Pais: Subtle Is The Lord, p 282.

difficult, but this is quite wrong. Einstein's contribution is very easy to understand, but unfortunately it rests on the theories of Galileo and Newton which are very difficult to understand!"[21]

Bondi is referring of course to the fact that both Galileo and Newton said that all motion is relative, but that we usually think of the place we are standing to be stationary. If we are going to encounter anything in special relativity that we might think is counter-intuitive, it will be this concept of the relativity of all motion. But I think that 'counter-intuitive' is the wrong word, since its popular meaning is close to 'illogical'. The concept of relative motion is just as logical as relative distance; we just don't deal with it as often in our daily lives. Maybe 'unfamiliar' would be a better word. With distance, relativity is assumed. Something can't be far away without being far away from something else. Yet we sometimes assume that something can be in motion all by itself, which, as Newton, Gallileo, Mach and Einstein have pointed out, can't be proven.

Inertial moving frames are the only ones to which special relativity applies, because it is only in these frames that all the laws of nature will apply in exactly the same way. Similarly it is only these inertial frames to which Galileo's 'principle of relativity'

---

21 James Gleick: Isaac Newton, p 194 (notes).

applies. If you pull out of a parking space quickly you feel the acceleration. But if the car beside you pulls out very quietly and gradually you may think your own car is rolling back. More than once I have hit the brakes and then felt foolish as the car beside me continued to pull out of its space. This is a good way to experience relative motion between inertial frames. When inertial frames are in motion relative to each other, we can't say that one is stationary and one is at rest.

It's hard to get this same feeling sitting on the front porch and watching the traffic go by. But it's the same thing, after all. The front porch and the steady stream of cars are in relative motion. A moment's reflection shows that this is common sense, not magic.

In the example of the space ship with the light bouncing up and down, acting as a clock, the ticks and tocks do look slow to the people on the ground, but this is not magic either. I like this simple but profound explanation from Lisa Randall better:

> "If I were to coordinate my watch with someone on a moving train, I would need to account for the time delay of a signal traveling between us because light has a finite speed."[22]

If we want a practical example, this is a good common sense one. If we check the watch of someone far away, we have to

---

22 Lisa Randall: Warped Passages, p 90.

wait for the light from that watch to get to us. If the distance stays the same, then once we start receiving the signal, the far away watch will tick at the same rate as our own. But if the distance to the second watch keeps increasing steadily, as on a train going farther and farther away, the time we see on the moving watch will tick more slowly and the time we read from it (if we could see that far) will fall farther and farther behind our own.

Now, has anyone's clock, 'really' slowed down, or does it just look like it has? If we accept the relativity of motion as put forward by Newton, Galileo, and Mach, then we can only answer the question as Einstein did; each person's clock really slows down for the other person. That is all we can say, politely, to a question that is flawed by the assumption of a fixed home plate. Asking which clock slows down is like asking which of us is far away.

Sometimes we tell our children things in a simplified way, thinking that the whole truth might be too much for them. There is a story about a stork... Sometimes, I think, scientists who are trying to explain special relativity to us, think that we wouldn't understand Galileo, Newton, and Mach, so they explain relativity in a way they think will make the most sense to us; as if the universe had a home plate and we were standing on it. This may explain why some say 'really' and some 'not really', each trying to give us the same message in a way that we will understand. I doubt there is much 'real' scientific disagreement on this. For our present purpose, once again let's listen to Einstein's own words:

> "… as judged from this reference body, the time which elapses between two strokes of the clock is not one second, but … a somewhat larger time."[23]

The phrase '… as judged from this reference body … ' says it all. So relativity of motion is about how things look in one inertial reference body or frame from another which is moving relative to it, and neither frame enjoys special status. If there is to be any magic, we will have to create it ourselves by imagining a fixed home plate.

The example of the spaceship with the two mirrors and the light going up and down between them is a good one for showing how the ticks and tocks can appear to the people back home to slow down. But I have two problems with this example. First, it encourages us to imagine who is moving, and who is stationary. Saying that one party flies off while the other 'stays' behind, makes us forget Newton, Galileo, and Mach. We think we are seeing clocks magically slow down, forgetting that they only slow down 'as judged from this reference body', to use Einstein's phrase again.

The other problem I have with the spaceship example is that it only works for me visually when the spaceship is moving away. When the spaceship is returning, my common sense expects

---

23 Albert Einstein: Relativity, p 42.

the ticks and tocks to be closer together, even though I know that they must still be farther apart since the light is still traveling both up and down and forward. Also, I have always wanted to ask what would happen to the ticks and tocks if you mounted the mirrors in the front and back of the spaceship, instead of on the ceiling and floor. As the ship moved away, instead of a regular, slow, tick … tock…tick…tock from the up and down light reflections, you would see an irregular …ticktock……..ticktock, from the light reflections going back and forth from the front and back of the ship as it moved away. After struggling with this and other visual models I finally came across a statement that put my mind at rest, from Abraham Pais:

> "Special relativity represents an abandonment of
> mechanical pictures, as an aid to the interpretation
> of electromagnetism."[24]

Thank you. Now I don't feel so bad for not understanding the mechanical pictures. They just don't seem to work very well. In Einstein's book 'Relativity', he introduces the mechanical picture of the train, the embankment, and the light beam, but immediately switches to some very simple math, such as we might use to calculate how many minutes it would take to drive home at a certain speed. Then he shows how the numbers change as the

---

24 Abraham Pais: Subtle Is The Lord, p 138.

reference frame changes, without having to embellish the mechanical picture.

A major source of confusion here is that we are dealing with two different concepts, and using the same word for both. Again our language isn't up to the job. We need to separate the 'principle of relativity' from the 'relativity of motion'. The 'principle of relativity' is the assumption, based only upon observation and common sense, that the laws of nature operate equally in all inertial frames. As long as the frames are inertial, this would apply with or without a fixed home plate. The 'relativity of motion' is an independent idea, which simply removes the home plate. Taken together, these two concepts mean that only motion is relative; everything else is invariant.

So, did Einstein say that moving clocks 'really' go slower than stationary ones? If we ask the question this way, we are still sitting in the theater after the curtain has come down and Jack has gone home. Or we are wandering through space looking for home plate, spike and hammer in hand. We don't want magic time to be over. We want someone to say 'really' or 'not really'. And no doubt, if we wait long enough, some kind physicist will come along and say one thing or the other, whichever they think will make us stop sobbing. When we get like this we need to go and sit in our car in the parking lot for a while, waiting for another car to pull out unexpectedly. When we have reclaimed the concept of relative motion, it is safe to go home and check the clock.

Chapter 6

CLOCKS

Our kitchen stove has a big digital time display on the upright part behind the elements. It is very handy. Just a glance from far away and you know that it is exactly '6:03', or '4:32', etc. You know the kind. But this one also has a built-in timer, and it can count down the number of minutes and seconds that you want, and when this amount of time is up the buzzer goes off. But the display does not go to '0:00' as you might expect. Instead it proclaims: "END" in big bright letters.

One afternoon I came downstairs from my nap, rubbing my eyes, and glanced at the clock. I had used the timer earlier and had forgotten to reset the display. So it was saying "END". Trying to make a joke, I said to my wife, who was already in the kitchen: "So that's what time it is." She glanced at the lit display and said: "Good. I've had enough." She was joking too, I hope.

Clocks don't know what time it is. They don't even know that they are clocks. Too bad. It would be nice to have a conscious clock friend, who could stick a finger into the flow of time, and tell us how fast time was moving today, how much time had gone before, how much was left, and whether there might be a way of persuading it to slow down.

The history of clocks is actually fascinating, and it was my original intention to cover the topic nicely right about here after doing a minimum of research. Forget about it. There is just too much good information out there, and it has already been presented better than I would ever be able to do. Not only that, but I found that the work already done goes way beyond the history of clocks; people have studied our relationship with the clock, and our changing concept of time itself. Apparently there are some societies who have no word for 'time'. Others have no words for 'future' and 'past'. Some speak of going forward into the past, and backward into the future. In our own culture our relationship with clocks and time keeps changing.

Even the length of our days and years keeps changing, due to the slowing down of the rotation of the earth. Our day gets longer by about 2.2 seconds every hundred thousand years. On December 31, 2008, one second was added to official clocks for this reason. A billion years ago an earth day was only about eighteen hours long.

So yes we have gone from sundials, to mechanical clocks, to quartz watches, and now to cesium atomic clocks which are accurate to about one second in twenty million years. But more importantly, our way of thinking about clocks, and what it is that they do, has itself changed and developed over the centuries.

Jo Ellen Barnett expresses this eloquently in her book "Time's Pendulum":

> "The invention of the clock began a gradual, centuries-long transition from a perception of time as something rooted in nature to something which originates in the clock itself."[25]

What an amazing statement, and what an amazing transition for us mentally. This explains how we can sometimes be persuaded today to make the leap from seeing a clock slow down, to thinking that time itself has slowed down.

But no, clocks don't know they are clocks, no matter how well they work, or how reverently we treat them. We can make a clock out of anything. We could use water dripping from the tap, or sand flowing through a narrow glass, or even a cow chewing on hay; any regular repeated action becomes a clock if we say so. All that is left for us to do is to give a name to the interval between the repetitions. The latest clocks use the repeated vibrations of an atom.

But the number of vibrations which would be called one second had to be decided upon by people whose job it is to do that sort of thing. The vibration of the atoms is more regular than the rotation of the earth. So somebody had to decide how many of

---

25 Jo Ellen Barnett: Time's Pendulum, p 145.

these regular vibrations would make up the interval called one second. They decided on a number which kept the convention of a 24 hour day, or at least the best approximation of it.

If the day gets longer they may decide to increase the size of the second. Then there would be more time between each tick and tock of the clock. This would not make time slow down. If we worked an eight hour day before they made the second longer, and kept working eight hours a day after the seconds were declared to be longer, we would not work in slow motion; we would work longer. Unfortunately, they would probably not pay us for the extra time.

So, just because my kitchen clock says it is the end, that doesn't mean it's over. And if the same clock were to slow down for some reason, or if a cow chewed slower, I wouldn't be persuaded that time itself had slowed down. When clocks run slow we fix them or throw them out. So when Einstein talks about a clock slowing down as judged from a different reference body, we need to ask ourselves if this means that anybody's time has slowed down.

Chapter 7

SLOW TIME

It can be fun to imagine what it would mean if time slowed down for some people but not for others. In my own imaginary scenario, if you and I are having coffee together and your time slows down for some reason, you will just disappear from my world. I don't know whether you went into my past or my future; all I know is that you aren't here now. Others might say that we will keep drinking coffee but you will drink more slowly, and age less than me while we are sitting together. Some people say that time is like a big loop, and we keep going around and around until we get it right. Others may imagine that there are many universes, following each other through time separated by split seconds. This would explain déjà vu, so the theory goes, because we may accidentally jump into the next universe for a second, and see into our own future. This is all in good fun to a point, or as Grandma said, 'till someone loses an eye'.

Indeed we can lose some sight doing this sort of thing. Mainly we can lose sight of Einstein's version of slow time. First of all, let's see how he defined 'time':

"Under these conditions we understand by the 'time' of an event the reading (position of hands) of that one of these clocks which is in the immediate vicinity (in space) of the event. In this manner a time-value is associated with every event which is essentially capable of observation."[26]

So we can feel safe to assign a time-value to an event by referring to the clock closest to the location of the event. This makes sense. Now what about a far-away clock, or a moving clock? Einstein has a clear, common sense answer to this also, and it is found on the same page of his book:

"It has been assumed that all these clocks go at the same rate if they are of identical construction." And furthermore, he adds: "... identical 'settings' are always simultaneous ... "

So there it is. Thank you very much Dr. Einstein. We are the ones who assign a time to an event. We use a clock close by. If we use several identical clocks close by, and then set some of them in motion in inertial frames, they would still be simultaneous when we brought them back, if we could indeed bring them back without changing their inertia. This is just what our common sense

---

26 Albert Einstein: Relativity, p 28.

wants to hear, and the best part is that we are hearing it first-hand from Einstein himself. Note that we would have to move them away and bring them back without subjecting them to changes in inertia, such as acceleration, or change of direction, so our example must remain theoretical.

On the very next page of this same book, Einstein begins a chapter titled: "The Relativity of Simultaneity". Here he reminds us that if we compare a clock by our side with an identical clock in motion relative to it, the readings will differ.. We don't get the clock reading from a distant frame until the light from that clock reaches our eye. Similarly clocks don't record events until the light from the event hits the clock. Even though our clocks may be identical and set simultaneously, they will not all receive light signals from an event simultaneously. They will not record the event at the same time. The only way they could receive the signal simultaneously would be if the speed of light were infinite.

And when two events are involved, the moving clocks may even disagree on which event happened first. No surprises here either. It is just a continuation of the same theme. No change in the clock settings. No change in the clocks' rate of speed. Things look different from different frames of reference, and notice of events reaches different clocks at different times.

Here is how Einstein said it:

"Every reference-body (co-ordinate system) has

its own particular time; unless we are told the reference-body to which the

statement of time refers, there is no meaning in the statement of the time of an event."[27]

In one of the Superman movies, Superman turns back time by flying so fast that time reverses, and a tragedy is undone. Conceivably Superman could find a frame of reference in which news of the tragedy has not yet arrived. Maybe on a planet far far away, like Superman's home. Abraham Pais puts it simply, in a short sentence which both states the concept, and emphasizes that there is no preferred frame of reference:

"There are as many times as there are inertial frames."[28]

In today's world we take for granted that everybody agrees on what time it is. We understand all about time zones and jet lag, and Greenwich mean time. This was not always the case. Not too long ago, when people used the sun directly to tell the time, the time could be different depending on which side of town you lived on. If you wanted to get to the blacksmith or the candle maker before closing time, you had to allow for the location of their shops relative to your house. There might be a few minutes

27 Albert Einstein: Relativity, p 31.
28 Abraham Pais: Subtle Is The Lord, p 141.

difference just across town. Eventually, with the advent of mechanical clocks, cities and towns started putting big clocks on towers in the middle of town. Then all the townsfolk could have the same time. But the next town down the road would have its own town clock, which might be different. Without telephone or e-mail, if you wanted to know what time it was in the next town, you would have to go there, or at least do some arithmetic. We still see these clock towers in towns and cities today, but we think of them as decorative now, and forget how important they used to be, and how independent they were of each other. The first trains had a hard time with scheduling. Here is how Lisa Randall described things as they were in Einstein's day:

> "Einstein worked in the patent office between 1902 and 1905, during an era when train travel was becoming increasingly important, and coordinating time was at the forefront of new technology. In the early 1900's Einstein was very likely thinking about real-world problems, such as how to coordinate the time at one train station with that at another." [29]

So when Einstein did refer to time slowing down, he was talking about clock readings. Time was defined by the readings on

---

29 Lisa Randall: Warped Passages, p 90.

clocks, and the readings varied with position and motion. If we do not get the same reading from two identical clocks in motion relative to each other, we can't say that time itself has changed. Theoretical physicist N. David Mermin put it this way:

> "While the notion that time stretches out for a moving clock has a certain intuitive appeal, it is important to recognize that what we are actually talking about has nothing to do with any overarching concept of time. It is simply a relation between two sets of clocks."[30]

Just as someone in Paris can figure out what time it is in New York, one can figure out the differences in clock readings between moving inertial frames. Even if events were observed in opposite order in different frames due to motion and proximity of the events, as long as one has the facts about the relative motion of the frames, everything can be transposed from one frame to the other. Physicist Robert Oerter reaffirms the common sense of it all:

> "Moreover, an observer who understands special relativity can easily change viewpoints, converting all his measurements into the reference frame of the other person. Doing so allows him to

---

30 N. David Mermin: It's About Time, p 63.

understand the others' conclusion about the order in which the events occurred."[31]

This is wonderful. Not only has our common sense agreed that things must look different in different moving frames because the speed of light is not infinite, but we can even sensibly 'change viewpoints', transforming and coordinating clock readings from all inertial frames regardless of their motion or location. But wait, I hear ominous music.

Apparently in 1972, two very accurate atomic clocks were synchronized, and one was flown at high speed around the earth, either on a plane or a satellite, depending on which report one reads. When the flying clock returned, it was slower than the one that stayed on earth, supposedly proving that special relativity was correct. But wait, there's more.

Not only is this experiment sometimes said to be proof of special relativity, but claims are also made that what has happened is that 'time' for the moving clock has slowed down, and this is what has made the clock run slow. This kind of thinking is what Jo Ellen Barnett was talking about when she said that we have come to think of time as something rooted in the clock. Now I don't doubt that the clock ran slow. But I do doubt that time itself slowed down, because if it did, then I am back in the coffee shop waiting for you to show up. The entire Universe as I know it exists

---

31 Robert Oerter: The Theory of Almost Everything, p 40.

right now, not some of it sooner and some of it later. If one clock had slipped backward in time, could we place the two side by side right now? Don't ask me. Also, didn't we just hear Einstein say that the clocks would always be synchronized, as long as they remained in inertial frames and unaffected by gravity or acceleration? And who, so far, has said that identical clocks in the same inertial frame, like on earth, would ever look different? As far as I can see, Einstein didn't say that. So what happened in 1972?

Looking for answers to what happens to clocks in motion, and the passage of time for objects in motion can be confusing. Here is what author Ben Bova had to say, about time aboard a (fictional) star ship traveling close to the speed of light:

> "Time itself is moving at a different rate aboard
> the speeding starship than it does for objects that
> are not traveling at relativistic velocities--such as
> the earth and everyone on it."[32]

So because of the 1972 experiment in which one clock ran slow after flying around the earth, we hear people claiming not only that moving clocks run slow, but that the reason they run slow is that time itself has slowed down due to the motion. But now my common sense is squirming and I am getting another headache. Are we back to the 'really' question, and has it been

---

32 Ben Bova: The Story Of Light, p 172.

proven that not only clocks, but also time itself 'really' slow down? As usual in this type of situation, I run back as quickly as possible, and as close as possible, to the source, to find out what Einstein said about this. Personally I am again relieved and comforted to hear the words of Einstein's colleague and friend Abraham Pais, who writes:

> "Einstein rather casually mentioned that if two synchronous clocks C1 and C2 are at the same initial position and if C2 leaves A and moves along a closed orbit, then upon return to A, C2 will run slower than C1. ....However, as Einstein himself explained sometime later, the logic of special relativity does not suffice for the explanation of this phenomenon (which has since so often been observed in the laboratory) since frames other than inertial ones come into play."[33]

'Frames other than inertial ones.' Meaning frames subjected to acceleration. Such as planes, or star ships. My headache is gone. What appears to have happened here is that someone has inadvertently mixed up special relativity, which deals with inertial frames, and general relativity, which deals with gravity. Maybe they also forgot that gravity and acceleration are equivalent.

Yes, general relativity says that gravity slows things down.

---

33 Abraham Pais: Subtle Is The Lord, p 145.

Things like light, and people, and clocks. So gravity's equivalent, acceleration, can do the same. One clock did slow down, but it was due to gravity and/or acceleration ('frames other than inertial ones'). It was not due to relative inertial motion. In short, it had nothing to do with special relativity. While I am glad that we have a reference to Einstein casually mentioning this, I sometimes I wish that he had not been quite so casual about it. But then I remind myself that he wasn't with us in 1972. So it looks like what Einstein did say about slow time is that moving inertial frames can have synchronized identical clocks, but since the speed of light is not infinite, when we move the clocks around differently we wait longer for signals from some than from others. No surprise. And since no reference frame gets special treatment, we can't say that one signal is more accurate than another. Even if we synchronize the watches together in one frame, and then separate them into moving frames, we can't prove that the first frame was not moving.

What Einstein did say was that synchronized clocks will read differently from different moving inertial frames, even though they remain synchronized. He also said that in 'other than inertial frames', meaning in frames influenced by gravity or acceleration or rotation, clocks will become inaccurate. What Einstein did not say, is that we can somehow slow down the flow of time by moving clocks around. If we are still looking for magic we will still be disappointed; all we have found, once again, is more good old common sense.

Chapter 8

THE TWINS

When I was a kid my friend told me that if you went to the moon and back really fast, when you got back you would be younger than your twin brother, if you had one. Well I didn't believe my friend, since his only proof was that Einstein had said it, and Einstein was a really smart guy. Anyway I had twin brothers, and they were the same age. This is the first instance in my memory when I thought that either this Einstein guy must have been wrong, or else he didn't really say that.

Years later I discovered that the twin thing had been brought forward as an argument against special relativity, not as an explanation of it. The claim was not just that one twin would be younger than his brother, but that they would both be younger than each other, which was impossible, thus proving that there must be something wrong with relativity theory.

Einstein apparently countered this by saying that the twin who had experienced acceleration would be younger, according to the clocks. So what we have here is just another example of the confusion between special and general relativity, the same as we had with the clocks in 1972. Yes special relativity means that a clock can look slow when being read from another inertial reference frame in relative motion. Yes general relativity means that a clock can be slowed down by forces acting on it. But once we separate the effects of special and general relativity, all confusion vanishes. Special relativity says that clocks look different from different inertial reference frames because the speed of light is not infinite, and we have to wait for light signals. General relativity says that in gravitational and accelerated frames physical processes, including the speed of light and the ticking of clocks, both slow down.

There have been so many arguments about these twins that they must both be getting old by now. All we need to do is remember we are dealing with clock readings, not time itself, and keep our accelerated frames separate from our inertial frames. Then we won't confuse the effects of special and general relativity. Yes there are effects; effects on perception between inertial frames in relative motion, and effects on physical processes under acceleration. Some people argue that when physical processes slow down due to acceleration, this includes the electromagnetic processes within our body which make us age. Maybe so, but when I am getting pushed back into my seat during takeoff on a 747, my knuckles go white and my heart beats faster. If I do this too often, I may age slower but die younger. I believe Stephen Hawking once wrote that the airline food would have the same effect.

And my brothers? Yes, they're still the same age as each other, and they're still getting older, even though they both travel quite a bit. So if I ever see my childhood friend again, I will break the news to him that Einstein didn't say that, even though he was a really smart guy. Actually my friend was smart too, so he has probably long since sorted this all out for himself.

Einstein Didn't Say That

Chapter 9

E=MC²

Now in case the gentle reader may expect that I am going to say that Einstein didn't even say that E=MC², let's get it over with. In 1905 Einstein wrote two papers on special relativity. They were not called special relativity. This name would come later, from other people. The first paper was called "On The Electrodynamics Of Moving Bodies", and it dealt with inertial frames in relative motion. The second was called: "Does The Inertia Of A Body Depend Upon Its Energy Content?". The famous equation comes from the second paper, but we might not recognize it right away. Here is how it was stated:

"If a body gives off the energy L in the form of radiation, its mass diminishes by L/C²."[34]

Here Einstein uses 'L' instead of 'E' for energy. So if we substitute the 'E' and use 'M' instead of the word 'mass', and then replace 'diminishes by' with an equal sign, we get: 'M=E/C²'. It was

---

34 Albert Einstein: Does The Inertia Of A Body Depend Upon Its Energy-Content? As reprinted in "A Stubbornly Persistent Illusion" by Stephen Hawking, p 32.

not until about two years after Einstein had published this paper that someone else made these changes, and then re-arranged the equation into its more familiar form. There is no change in meaning, mathematically. It's like saying that 6 equals two times three, instead of saying that two equals six divided by three. So, while you don't have to worry about me claiming that Einstein never said that $E=MC^2$, it is interesting, nevertheless, that he did not say it this way the first time, and that he was not the first person to say it this way. If anything, this may show that his attention was on the mass, so he put the 'M' first. Consider this, from Einstein:

> "The most important result of a general character to which the special theory of relativity has led is concerned with the conception of mass. Before the advent of relativity, physics recognized two conservation laws of fundamental importance, namely, the law of the conservation of energy and the law of the conservation of mass; ...By means of the theory of relativity they have been united into one law."[35]

Sounds easy. Why didn't somebody think of this before? The fact is that while this was the outcome, it wasn't something

---

35 Albert Einstein, Relativity: p 51.

that Einstein knew he would find. William Cropper says that originally Einstein was just exploring the implications of his two assumptions:

> "Relativity begins with a modest question: How does your physics relate to my physics if we are moving relative to each other?"[36]

This is the question asked and answered in the first 1905 paper about moving bodies. When it was thought that the speed of light was infinite, motion would not affect how things looked when they moved. But assuming that the speed of light is constant, and less than infinite, changes the way things look when they move. Measurements of length come up a tiny bit short, because the light has traveled farther from one end of the object than it has from the other. Light signals received simultaneously from the front and rear of a moving object were actually sent at different times; the signal from the front was sent a tiny bit earlier, when it was a tiny bit further back, making the observer's measurement of the whole object come up a tiny bit short. The amount of difference is related to the speed of light.

The second paper goes further, saying that even sitting still, matter changes; it gives off radiation, and in the process becomes a tiny bit smaller. Michelangelo knew that he would find

---

36 William H. Cropper: Great Physicists, p 201.

David in the stone. But Einstein was just chipping away. Beginning with trying to make sense of what was already known about the electromagnetic structure of light and the principle of relativity, he ended up focusing in on the nature of mass. And it turned out that mass was energy. Mathematician Robyn Arianhrod assures us non-mathematicians that it is ok to drop the C^2 from the equation:

> " ... Einstein's famous $E=mc^2$, ... can be rescaled to give $E=m$, an equation whose unearthly consequence is that ethereal, immaterial energy (E) is equivalent to --interchangeable with--solid matter (m)".[37]

It makes one wonder why it isn't written this way more often. Less exotic, but easier: energy equals mass.

---

37 Robyn Arianhrod: Einstein's heroes, p 4.

Chapter 10

MASS

Physicist Hans Christian Von Baeyer looks at how Einstein regarded M, or mass:

> "Mass, denoted by m, is a measure of inertia--the tendency of things to resist being set in motion when they are at rest and to resist changes in velocity when they are moving....To the physicist, inertia and mass are expressions of the same property, one qualitative and the other quantitative. Indeed, Einstein chose to use the qualitative word in the title of his paper "Does the Inertia of a Body Depend Upon Its Energy Content?" and not to mention mass at all."[38]

By explaining this distinction between the 'qualitative' term 'inertia', and the 'quantitative' term 'mass', Von Baeyer takes a step toward clearing up another huge popular misconception about what Einstein didn't say. Just as we have all heard that gravity is not

---

38 Hans Christian Von Baeyer: Maxwell's Demon, p 124.

a force, and that time slows down, we have also heard that mass increases with velocity. If only those who write some of this popular titillating stuff would read folks like Von Baeyer, fewer of us would believe we have to choose between common sense and science.

The reasoning behind the increasing mass idea, presumably, is that moving objects have more kinetic energy than stationary objects, and since energy is equal to mass, then adding kinetic energy also adds to the object's mass. The familiar problem here is that somebody keeps forgetting that motion is relative. Somebody needs to go back and revisit Newton, Galileo, and Mach. Otherwise we are back asking the 'really' question. How can you say which of us is really in motion? Which of us gains the weight when our inertial frames are in relative motion?

The resistance of objects to changes in velocity, that is, the 'inertia', of objects, doesn't change due to motion. The 'MC²' expression doesn't include a term for motion. Nor does it include a term for velocity of such motion, or kinetic energy, or heat energy, or nervous energy, for that matter. If we want an equation that includes these things we will have to order it from the math shop on the corner. We have confirmation from Einstein:

> "The first term mc² does not contain the velocity
> ... "[39]

---

39 Albert Einstein: Relativity, p 50-51.

The caveat here is that in the process of identifying what Einstein meant by 'M', we don't want to oversimplify the concept of inertia too much, as it is still under study from lots of angles. In 1982 Abraham Pais wrote:

> "It must also be said that the origin of inertia is and remains the most obscure subject in the theory of particles and fields. Mach's principle may therefore have a future--but not without quantum theory."[40]

And today, almost thirty years after Pais wrote this, scientists are still attempting to isolate the subatomic particle that could be responsible for gravitational and inertial mass, by building huge machines to smash up matter. But for our present purpose we only want to know what Einstein meant by the term 'inertia', in 1905, and to remember that, like Newton, he didn't claim to know the 'origin' of inertia. What inertia is, we know. What causes inertia, we don't yet know. A physicist whom I once met had a sign on his office wall which read: "If we knew what we were doing, it wouldn't be research." Or as much fun, presumably. And yet things get done. Advances are made.

There was one advance in the knowledge of the nature of mass that would be crucial to special relativity, and it came just in

---

40 Abraham Pais: Subtle Is The Lord, p 288.

time for Einstein to put it to use. This advance was made by a Dutchman named Hendrick Lorentz.

Chapter 11

LORENTZ

Of special interest to Einstein was the fact that Lorentz had treated physical objects as electromagnetic in their structure. Einstein gives Lorentz a generous amount of credit for this as a contribution to relativity theory:

> "Nevertheless we must now draw attention to the fact that a theory of this phenomenon was given by H.A. Lorentz long before the statement of the theory of relativity. This theory was of a purely electrodynamical nature, and was obtained by the use of particular hypotheses as to the electromagnetic structure of matter."[41]

Einstein's first 1905 paper on the subject was called "On The Electrodynamics Of Moving Bodies". Perhaps this title could have been used for some of Lorentz' work. When we hear in the above quote the words: 'The electromagnetic structure of matter', it may seem like a bland- sounding phrase, but in fact it describes an astounding reality. Once during an open house at our local

---

41 Albert Einstein: Relativity, p 46.

university, I noticed a young physics student standing beside a magnetism display, and I took the opportunity to ask him something that had always made me curious:

"Why don't those little fridge magnets eventually fall off the fridge?" I asked. It seemed to me that these ornamental little gadgets must be using up some energy, hanging on for dear life like that, and holding up not only themselves, but notes and reminders of varying importance all the while. Wouldn't they eventually get too tired and weak to hang on? I know I would. Why didn't they run out of magnetic energy?

The young student gave me a great answer. He held up an object, a ruler I think, and asked me why it stayed intact. What was holding it together? He then told me that whatever it was that was holding this object together, was the same force that was keeping the magnets on the door. Suddenly I realized that the magnets didn't think of themselves as hanging on to the fridge at all; they thought that they were part of the door. They weren't about to fall off the door, any more than the door itself was about to fall apart.

Thanks, kid; this is beautiful. Maxwell and others had described the electromagnetic nature of light waves, but by describing the electromagnetic structure of matter, Lorentz set the stage for Einstein to discover the equivalence of mass and energy. Any changes taking place within the structure of matter would happen at the speed of electromagnetic activity; that is, at the speed of light. So a non-mathematician like myself is able to glean from this the first notion of why the speed of light might be involved in the relationship between mass and energy.

Einstein's first statement of this relationship, as we have seen, talks about matter giving off energy in the form of radiation. The mass has shrunk by an amount related to the speed of radiation; the speed of light. Determining the amount of radiation a certain amount of matter could give off would require a mathematical formula. And Lorentz had come up with just such a formula.

Einstein Didn't Say That

Chapter 12

CONTRAPTIONS

The formula Lorentz had come up with is referred to as the 'Lorentz Transformation', or the 'Lorentz Contraction', or the 'Lorentz transformation contraction formula', or just the 'Lorentz equations', depending on the source one consults. It is a set of equations which measure how much something contracts, or shrinks, for whatever reason. It could probably be applied to what happens to sweaters in a clothes dryer. The bigger the sweater the more it shrinks. Since to someone like me the Lorentz Contraction is a complicated looking bit of machinery that does a straightforward job, I like to call it a contraption. The 'Lorentz Contraption', is my term for it, coined with no disrespect. Giving it this name just means that I don't have to worry about how it works, only what it does. Kind of like a clothes dryer, but more reliable and cheaper to operate.

Sometimes contraptions can be used for purposes other than those intended by the inventor. Once a group of us were talking about the pros and cons of being self-employed, and it was suggested that if one were self employed, lots of expenses could be used as income tax deductions. Even if certain expenditures were not directly related to one's business, a receipt sometimes got put in anyway. One fellow admitted to having 'put in' his child's new tricycle as a starter motor for his taxicab. There was a brief silence before someone astutely asked: "Did it work?"

Lorentz's equations were designed to measure the contraction of objects due to the resistance they experienced while traveling through the ether. The formulas were based on earlier scientific work, which would also have been known to Einstein.

In 1842, a scientist by the name of George Gabriel Stokes had come up with a formula for finding out how much kinetic energy a ball would lose when traveling through a liquid. Not surprisingly, the energy loss depended on the radius of the sphere, among other things. The bigger the ball, the more kinetic energy it would lose as it moved through the liquid. Every kid who has played with a beach ball in the water knows about this. The difference is that Stokes did the math. So the mathematics for figuring out energy lost by a sphere in motion was well established before Einstein's time.

Lorentz had adapted Stokes' work to objects moving in a presumed ether; he wanted to measure the amount of shrinkage things would experience, due to compression, as they moved through the ether. By doing this he hoped to explain the results of the Michelson and Morley experiment. He figured that if their equipment had shrunk due to motion, this could explain the unexpectedly consistent measurements they got for the speed of light. The light could have been slowed down by the ether, but would still arrive on time if the experimenters' apparatus had contracted. The shorter length would exactly compensate for the slower speed. Today many biographers refer to Einstein's use of Lorentz' equations in special relativity. One might think that Einstein just plugged the Lorentz contraption into a wall socket and it spit out E=MC². In fact Einstein himself seemed to allow, if not encourage, this idea, referring often to the 'Lorentz Contractions'. But it is interesting to note that Abraham Pais doubts that Einstein had access to this work by Lorentz in 1905:

> "As to Einstein himself...he repeatedly pointed out elsewhere, in 1905 he knew Lorentz' work only up to 1895. It follows-- as we shall see-- that in 1905 Einstein did not know of the Lorentz transformations. He invented them himself."[42]

Whether Einstein adapted, or re-invented the formulas, he did use them. My own impression is that when Einstein repeatedly mentioned the contributions of Lorentz, he was referring less to the equations themselves, than to the very crucial, fundamental concept of the electromagnetic structure of matter; the fridge magnet thing.

Some of us non-mathematicians may not be able to

---

42 Abraham Pais: Subtle Is The Lord, p 121.

completely understand an equation such as $E=MC^2$, or how it works, but that doesn't mean we can't try to recognize the component pieces and understand their purpose. Sort of like trying to assemble a tricycle which came in pieces in a box, without reading the instructions. In fact many men from my own generation find it a little annoying when someone suggests that we read the instructions; we find it far more sensible to just lay out the parts on the floor, and try to identify what they are and why we might need them. The inventor in each of us is secretly expecting to find a new and better way to put the thing together. In the case of trying to put together the parts of $E=MC^2$, we have so far laid out E, and M, and C. We all know what they should look like after they are assembled. But we may not yet see why they must go this way. And in my case, my common sense is still asking for a better idea of what exactly the C is doing, stuck on the end like that, and why it has to be squared.

Chapter 13

C^2

Several times I have heard physicists explain the C^2 factor by saying that it is important because it is such a 'big number'. Well, without wanting to sound at all ungrateful to the same physicists who have given me so many hours of pleasurable contemplation, and from whom I admit I have absorbed only a small portion of what they know, the 'big number' explanation in this case just isn't quite enough for me. First, if C is a constant, why do we need it at all? If the equation means that matter can be converted to energy, and vice versa, why not just say E=M, and go home early?

The other thing about all this is that it seems to me that the very reason C^2 is a big number is because we use such small units. About three hundred thousand kilometers per second, multiplied by itself. But what if, instead of measuring the speed of light in kilometers per second, we measured it in light years per year? Since light travels a distance of one light year in one year, then the number for C would be one, and so would the number for C^2. One times one is one. So now we would multiply mass times one, instead of ninety billion, or whatever it works out to the other way.

My question is not so much about how big it is, but rather, why we need it. Surely there must be a more meaningful way to think about C^2. Not that I ever doubted that C^2 represented a large quantity; I guess I just wanted to better understand how the number we used for the speed of light related to the numbers we used for mass, and why it had to be squared. It was like staring blankly at one of the pieces of the tricycle on the floor, knowing it must be important, but not knowing why. When I finally did see how it fit, and why it had to be squared, it was like the moment when we see how the tricycle parts are going to fit together, and can't wait to demonstrate our genius to the doubting children.

Usually we think of light waves as traveling in one direction, as they do from a flashlight. But a single spark does not form a beam and head in one direction. Rather it goes in all directions, in the shape of…yes, …a sphere. There it is. We are back to Stokes and Lorentz. Spheres and formulas. Beach balls and tricycles. It's time to call the children; things are going to fit together. How big is the sphere? Well, it gets bigger over time, at the speed of light. Since the speed of light is limited, so is the size of the sphere. The radius of this sphere depends on the speed of light. So now we can easily make the mental connection between C and the beach ball. The C will be used in the contraption that measures what happens to the ball due to motion. It will give both the radius and the maximum rate of expansion of the sphere. The more mass there is, the longer it will give off light, and the bigger the sphere will become. Einstein packs all of this into one concise statement, while at the same time reminding us yet again of the relative nature of motion:

"The wave under consideration is therefore no less a spherical wave with the velocity of propagation c when viewed in the moving system."[43]

This is a picture that makes sense, and so far without using any numbers at all, big or small. We get what the C is used for. We can relate it all the way back to the work of Stokes in 1847, or beach balls in the lake. But which number should we use for this C of ours, and why must we square it?

Many of us who don't know much math, will at least remember from our schooldays that the surface area of a circle has something to do with the square of the radius. So we won't be surprised to discover that the formula for finding the surface area of a sphere also involves the radius squared.

George Stokes and Hendrick Lorenz had spheres losing energy, or being compressed due resistance to motion. They used radius 'R' squared in their calculation of the surface area of the sphere. Einstein's sphere had a radius determined by C, so he used C^2. But instead of saying that measurements would decrease due

43 Albert Einstein: On The Electrodynamics Of Moving Bodies. As reprinted in 'A Stubbornly Persistent Illusion', by Stephen Hawking.

to compression of the electromagnetic structure of the matter, Einstein said that the contraction was due to the conversion of mass to energy, through electromagnetic radiation. In either case the radius of our beach ball depends on the speed of light. Using the Lorentz formula, and C to find the radius, Einstein could measure the amount of mass being lost due to radiation.

When he said, in his 1905 paper, that "mass diminishes by L/C^2", this represented the amount of mass that got converted to energy. When the equation is written $E=MC^2$, the E represents total amount of energy which could be radiated, theoretically, from a given amount of mass. It is the total amount of energy which will be radiated from the beach ball in its inertial state, by the time the whole beach ball is entirely converted to energy. The confirmation of the accuracy of the formula didn't come along until people began experimenting with nuclear reactions on a large scale:

> "In 1937 it was possible to calculate the speed of light from nuclear reactions in which the masses of the initial and final products and also the energy release in the reaction were known. The resulting value for c was accurate to within less than one half of one per cent."[44]

---

44 Abraham Pais: Subtle Is The Lord, p 149.

Now here is the last sentence from Einstein's paper, written 32 years earlier, in 1905:

"If the theory corresponds to the facts, radiation conveys inertia between the emitting and absorbing bodies."[45]

Mass getting converted to energy, and back again, according to a formula. Much as some of us may regret our inability to clearly understand the formula, we are nevertheless able to get the main idea. George Stokes had experimented with solid spheres, and Maxwell and Lorentz had talked about the electromagnetic composition of light and matter respectively. Then Einstein adapted the Lorentz contraption to measure loss due to radiation, rather than compression, and showed that the amount of energy in a small chunk of matter comes out equal to the light energy in a very big sphere.

And so we can use whatever numbers we want for the speed of light; the radius of the sphere, and the square of that radius, will be the same physical size whichever numbers we use. Which brings us to the next question: can numbers be used for more than just measuring, or counting things?

---

45 As reprinted in Stephen Hawking's: "A Stubbornly Persistent Illusion". p 34.

Einstein Didn't Say That

Chapter 14

MATHEMATICS

There is an interesting book called "Zero: the biography of a dangerous idea", by Charles Seife, which shows a series of simple equations that can be used to prove that Winston Churchill is a carrot. The reference to Churchill may be a little dated, but I'm sure that the formula would work just as well to turn more current politicians into carrots. Not a bad idea, some might say. Unfortunately the proof involves dividing by zero, which they tell me is against the rules.

And speaking of rules, who makes them? Who gets to say that you can't divide by zero, besides the politicians themselves, I mean? And while we're at it, what makes imaginary numbers ok? Why is it ok to use the square root of negative one, even though negative numbers don't have square roots? From what I have read this used to be against the rules, but the rules got changed. Did a law of nature change, or did our standards change, or did we just get smarter and realize our previous error? Don't ask me. Now they use the square root of negative one, and call it an 'imaginary number'. Of course there's no way I'm going to question them; they should know.

My very smart four year old grand-nephew decided that he wanted to play the cello. First his parents made him a very nice toy model, and then they bought him a real one, appropriately sized. Recently he told someone that he has two cellos. One of them, he said, is a real one, while the other one is imaginary. I thought it was interesting that he used the word 'imaginary', where many four-year-old's might have said 'pretend'. Obviously the mathematicians and my grand-nephew all know something that I don't, about the difference in meaning between 'imaginary', and 'pretend'. They wouldn't use pretend numbers, would they? But then I said I wouldn't question them.

One person who didn't like to use the square root of a negative number was Rene Descartes, who is famous for saying: "I think therefore I am". What he thought about a lot of the time, when he wasn't thinking about whether he existed or not, was mathematics, and he actually invented some of it. He also invented the name 'imaginary number', for the square root of negative one, since he thought that the whole concept was 'fictitious'.[46]

Sometimes you can't make things go away just because they don't fit the name you have ready for them. The word 'number' doesn't really work for certain concepts, so we qualify it with adjectives. 'Irrational numbers', which are those ones with decimals that go on forever, aren't really numbers in the sense that you could use them for counting. You can cut an apple in thirds and count

---

46 See "The Great Equations." by Robert P Crease, pg 97.

the three pieces, but if you try to measure out a third by counting decimal places you will never finish, because the 'number' for one-third, expressed in decimals, goes on forever. Similarly the number for pi also goes on forever, and isn't much good for counting things. It expresses a relationship between parts of a circle, rather than an amount or size of anything. And of course everyone is familiar with the old puzzle about going halfway to someplace, then halfway again, and then again; you never get there. But these concepts are no less real, just because our counting numbers don't fit. Some ideas that we can imagine just can't be described using the same numbers that we use to count things. And some people might argue that even the counting numbers are imaginary, since they are really just concepts that exist only in our minds. You can't see or touch or taste 'seven', all by itself. If you don't break any rules, your mathematics should be correct. Even if the numbers represent relationships or concepts rather than countable objects, the math can be correct. But just because the math is correct, that doesn't necessarily mean that it corresponds to something real. We can double seven to get fourteen, but that doesn't mean we have fourteen of anything real. Physicist Lisa Randall says it better:

> "A mathematical theory must be internally consistent but, unlike a scientific theory, it has no obligation to correspond to an external physical reality."[47]

---

47 Lisa Randall: Warped Passages, p 104.

This is something we don't hear often enough. More and more products these days have warning labels; maybe this warning should be on all mathematical theories: 'Guaranteed to be correct, but may not correspond to reality'. Einstein expressed a similar thought near the beginning of his book on relativity, in reference to geometry:

> "... geometry, however, is not concerned with the relation of the ideas involved in it to objects of experience, but only with the logical connection of these ideas among themselves."[48]

So both Randall and Einstein have said that your mathematics can be correct without being connected with reality. Writing this so early in his book, Einstein seems to be setting the stage to say that common geometrical ideas like straight lines are going to be difficult to apply in some cases. Kind of a polite way of telling Euclid that while his straight lines are nice, we won't be needing them much when describing how planets & light beams move in space when they are subjected to gravity.

So in this case we could say that Einstein was rejecting one form of math in a particular situation, not because it broke any rules, but rather because it wouldn't correspond to the curved trajectories of light and objects as they moved through

---

48 Albert Einstein: Relativity, p 4.

gravitational fields. The math had to fit the reality. Rather than needing less math, Einstein was going to need to learn more of it, specifically that which could deal with curved lines instead of straight ones. Biographer Walter Isaacson passes on this quote from Einstein:

> "At a very early age, I made an assumption that a successful physicist only needs to know elementary mathematics. At a later time, with great regret, I realized that the assumption of mine was completely wrong."[49]

In the early years Einstein used mathematics to record the flashes of insight brought on by imagining himself riding bicycles at the speed of light, or falling from chairs. But later on he seemed to be trying to find new ideas just within the mathematics. Einstein began working with the mathematics to find a 'unified field theory'. Everyone joined in. Physics, thanks at least in part to Einstein, the guy who had earlier thought that mathematics wasn't that important, was becoming more mathematically oriented.

As a result we now have many brilliant, mathematically-driven theories. How do we know if they are right? In his book "The Trouble With Physics", physicist Lee Smolin writes about the amount of effort that has been put into creating beautiful theories

---

[49]Einstein, as quoted by: Walter Isaacson: Einstein, His Life and Universe, p 33.

using mathematics, and the lack of experimental success in proving their connection with reality:

> "It is not an exaggeration to say that hundreds of careers and hundreds of millions of dollars have been spent in the last thirty years in the search for signs of grand unification, super symmetry, and higher dimensions. Despite these efforts, no evidence for any of these hypotheses has turned up."[50]

So what should we expect from mathematics? Is it just a language? Is it just a way of counting things, like income tax deductions and tricycles? That doesn't seem to quite cover it. Even at the elementary level we can appreciate the truth and beauty of a mathematical idea. How do we balance the inspirational with the observational? Let's have a look at what some people have said, beginning with mathematician Robyn Arianhrod:

> " ... mathematics is a language for thinking dramatic new thoughts, not merely for doing accurate bookkeeping."[51]

And physicist Janna Levin goes a little further:

---

50 Lee Smolin: The Trouble with Physics, p 176.
51 Robyn Arianhrod: Einstein's Heroes, p 96.

"Mathematics can penetrate into realms where our eyes fail us and three dimensionality blinds us."[52]

Next, Einstein biographer Robert G. Newton:

"If Galileo thought that mathematics was the language of nature, some physicists now go even further, believing that the laws of nature should be mathematical theorems."[53]

And finally at the pinnacle of this continuum, here is Hertz, being quoted by physicist Frank Wilczek:

"One cannot escape the feeling that these mathematical formulae have an independent existence and an intelligence of their own, that they are wiser than we are, wiser even than their discoverers, that we get more out of them than was originally put into them."[54]

How far does one go with this? If a mathematical theorem doesn't break any rules, like dividing by zero, is it in some sense a

---

52 Janna Levin: How The Universe Got Its Spots, p.104.
53 Robert G. Newton: From Clockwork to Crapshoot, p 3.
54 Heinrich Hertz, as quoted by Frank Wilczek in "The Lightness of Being", p 103.

law of nature? What are the limits to the power of mathematics? Here to remind us that there are in fact limits, is Janna Levin again. This time she is discussing the work of Godel on the provability of mathematical theorems:

> "Devastatingly, Godel managed to prove that there are propositions which cannot be proven or disproven within the context of axiomatic mathematics. He proved that some propositions are unprovable. This isn't to say that mathematics is rendered useless but just that mathematics must not be entirely self-contained and not entirely comprehensive."[55]

Thankfully Levin has brought us back to solid ground. It doesn't seem devastating at all to find out that mathematics, at least when we are talking about physics, should connect with something outside itself. This takes us back to what Einstein said about geometry. Mathematics for its own sake still works, and can still be worthwhile, but sometimes it looks more like art than science. So while we can be thankful for the new ideas we get when smart people fill computer screens with numbers and symbols, we can also reserve the right to hit the 'enter' button once in a while to see if the bottom line represents something pertaining to physics, as in

---

55 Janna Levin: How The Universe Got Its Spots, p. 190.

some physical object or event.   Of course we will still have to take the word of the mathematicians about what the rules are. We will avoid dividing by zero, because they have shown us the unpleasant consequences of doing so, and we'll not ask the difference between the adjectives 'imaginary', 'pretend', and 'fictitious' when they are applied to the   number which is said to be the square root of negative one.

Personally I can't help wondering whether this imaginary but legitimate thing of many modifiers could be one possible gateway between the correct mathematics which corresponds to reality, and the correct mathematics which is just correct mathematics.   It does seem that there must be a gateway somewhere in the math, which leads to places filled with more beauty than truth. But before we look for this gateway, perhaps we could take a quick look back at how we got this far.

Einstein Didn't Say That

Chapter 15

REWIND

In our empty nest we have acquired certain comfortable habits which allow us to set aside the business of life, when possible, and reflect on something. The subject matter being reflected upon is less important than the act of reflection itself. Comedian Lewis Black once said that if a place is suffering economically, then what it should do is build a big *** thing. Then everyone will come to see this big *** thing, and while they are there they will spend money in the big *** thing restaurant and then spend some more money in the big *** thing gift shop. "I don't care what it is.", he said, "As long as it's big, and it's a *** thing." People like to reflect on things.

In our case the benefit of taking time to reflect on something is that it reminds us that there are parts of life which are more important than what happens. Often this moment arrives about halfway through a movie we are watching. I will turn to her and ask: "What's this about?" The answer invariably comes back: "I knew you were going to say that." Sometimes that is the end of it, and sometimes that is enough, and sometimes we surprise ourselves and find something new to say about life. Sometimes we actually try and figure out what the movie is about, and even that is enough. Meanwhile of course we have missed part of the movie and have to play some of it back.

With the gentle reader's permission, I would like to rewind here just a bit, and highlight what I think are the most important things we have seen so far, that Einstein did and didn't say. There's no way we could track down all the things Einstein didn't say and prove that he didn't say them. All one needs to do is type the words "According to Einstein … " into an internet search engine in order to see the scope of the problem. But in my opinion one of the main things that Einstein didn't say, is that time slows down for moving clocks.

Einstein started out by trying to reconcile the assumption that light always has the same less-than infinite speed with the assumption that the laws of nature are the same in all inertial frames. This means that we can't measure our own speed against the speed of light, since we always get the same answer. This also means that we can't determine whether or not something is in motion, just by comparing its speed to the speed of light. In fact we can't measure the motion of anything against the speed of light, and we can't even say for sure whether or not something is in motion. All we can measure, and all we can prove, is whether or not something or somebody is getting closer to, or farther from, something or somebody else.

Einstein described a thought experiment in which there is a man on a train, and a man on the train station platform. The train is in motion relative to the platform. There is a beam of light in motion relative to the train and the platform, and the light is going in the same direction as the train. The man on the train has a watch, and the man on the platform tries to read that watch. The two men are getting farther apart, so light will take more and more time to travel between them. If light speed were infinite, there would be no waiting time. But since the speed of light is less than infinite, the man on the platform sees the watch ticking more slowly as the train moves away. With each tick and tock the light from the watch must travel farther before the person at the station sees it. 'Time' is defined as the reading of watches or clocks, taken when the light from the watches or clocks reaches the person reading them. As the watch on the train and the person at the station get farther apart, the person at the station has to wait longer and longer for the light from the watch to arrive. Here is part of Einstein's description:

> "Now let us suppose that our railway carriage is again traveling along the railway lines with the velocity v, and that its direction is the same as that of the ray of light ... ".[56]

The train's velocity relative to the platform is unchanging; it is 'v'. Because the velocity is unchanging, the inertia is constant,

---

56Albert Einstein: Relativity, p 22.

which only means that the train is not accelerating or going around a bend. There are two people in two inertial frames, moving steadily apart. Einstein's thought experiment involving the train does not involve time slowing down, or any other form of magic. It only involves somebody waiting for a light signal from somebody else's watch. It is easy to understand.

On the other hand, the example of the spaceship with the beam of light bouncing up and down between two mirrors leaves open a dangerous opportunity for misinterpretation, in my opinion. In this example, we are asked to imagine that the time it takes for the light to travel from the ceiling to the floor of the spaceship is one second, and that the light takes another second to bounce back to the ceiling. This bouncing light beam is in effect a clock, which counts seconds. No problem so far. A clock can be made from any regular repetitive pattern.

But then we are told that when the spaceship is in motion, the light beam travels farther in one second than it did when the spaceship was standing still, since now we have added forward motion to the up and down motion, making the light travel farther with each bounce. This is where the alarm bells should ring in our ears. We have been asked to forget that motion is relative, by choosing a stationary inertial frame (the earth), and a moving inertial frame (the spaceship), rather than two inertial frames in relative motion. Once we fall into this hazard, it's hard to escape.

It is only when we ignore the relativity of motion that we can accept the next step on this road to misinterpretation. The next step goes as follows: If the speed of light is fixed, and yet light travels farther in one second than it is supposed to, then something has to give. The seconds must be longer for the moving spaceship. And then finally, the fateful conclusion is that time slows down for moving clocks. Here are the missteps again. Each one leads to the next:

Step 1: The spaceship moves, the earth stands still.

Step 2: Therefore, motion is not relative.

Step 3: With motion, each bounce is longer.

Step 4: With motion, each second is longer.

Step 5: With motion, each second is slower.

Step 6: With motion, time is slower.

Step 7: Time slows down for moving clocks.

By concentrating on the distance the light travels between the mirrors inside the spaceship, instead of the distance from each mirror to the person at the station, we have thrown the principle of relativity out the window. We have assumed that the earth is stationary and the spaceship is not. This is the mistake that leads to so many more. We cannot prove that the spaceship is moving and the earth is not. We can only say that they are getting farther apart. It would be at least a little less confusing if we just had a single flashing light on the back of the spaceship. The measurements of the time taken for light to reach the earth would be the same as with the bouncing light, but those measurements would be easier to understand. We would just end up with somebody waiting for the flashes of light to arrive, like the person at the train station. In the train example we can attribute all the motion to the train, or all the motion to the station, or some to each. It doesn't matter which. All that matters is the increasing distance between the two people.

Now my guess is that when certain physicists tell us that time slows down for moving clocks, it is because they are trying to describe to non-mathematicians like me how the math works from here on earth where we live. They want us to understand why the amount of time represented by 't' in Einstein's mathematics increases. Yes it does. But the 't' in Einstein's math marks only the arrival of light signals from the watch being received by the man on the platform. Common sense tells us that the time between the arrivals of these signals increases as the light travels farther.

Nobody's time has slowed down. If I expect you home in an hour, and you arrive late, I won't think that your time slowed down. I will realize that your arrival time depends on our relative position, relative motion, and the velocity of that motion. The problem with approaching Einstein through the math, it seems to me, is that one just might forget the there needs to be a hook to reality. If we try to understand what is happening by only looking at the math, we might be tempted to attribute more meaning to that little 't' than it deserves. Here's Einstein again, reminding us to keep a physical measuring stick hooked to our mathematics:

> "Science has taken over from pre-scientific thought the concepts space, time, and material object (with the important special case "solid body"), and has modified them and rendered them more precise. Its first significant accomplishment was the development of Euclidian geometry, whose axiomatic formulation must not be allowed to blind us to its empirical origin (the possibilities of laying out or juxtaposing solid bodies).[57]

So in the end, after all the math is done, we have to take it outside and try it out. Of course Einstein did use math himself, to explain, explore and expand upon the concepts which he had introduced with the guy on the train, and he did discover some amazing new stuff in this way. But that stuff turned out to be measurable. It turned out to be real. Like the amount of energy in matter.

The same effect of waiting for light that made clock readings change with relative motion, also makes physical measurements change, and for the same reason. If we stand on the railway platform and watch a train car pass by, then at the instant

---

57 Albert Einstein: Relativity, p 162-163.

the back of the car passes us, we wait longer for the light from the front end of the car to reach us than we do for the light from the back end. The front is farther away from us at this instant. If we take both the front and back measurements at the instant the back of the car passes us, the signal from the front is inaccurate.

The light reaching us from the front of the car at the instant the back of the car passes us, was given off a tiny bit earlier. A bit earlier the car was a bit farther back. Our measurement comes up a tiny bit short. Now all of this could have amounted to no more than a curiosity; an observation of the same variety as noticing that things look smaller when they are far away. One might reasonably ask: "So what?"

Well, what if a real thing could really become smaller, by giving off energy in the form of radiation? How much shrinkage would there be? The answer is that the shrinkage due to radiation is the same as the apparent shrinkage we saw due to relative motion. In either case the same formula, the Lorentz transformation, would explain the amount of shrinkage.

The Lorentz transformation was no ordinary shrinkage formula. The shrinkage factor in this particular formula was based on the speed of light. The reason it was based on the speed of light was because Lorentz discovered, or realized, or theorized, that the structure of matter was electromagnetic.

And due to the electromagnetic structure of material objects, like refrigerators and things that stick to them, the amount of radiation energy contained in matter is also based on the speed of light. When you take all motion out of the shrinkage formula, you get $E=MC^2$. We all know the good ways and the bad ways that this equation is hooked to reality. I'm guessing that Mr. Black would agree that this qualifies as a big thing, well worth reflecting upon.

The other main things that I think we have found so far, that Einstein didn't say, had to do with gravity. Einstein leaned back in his chair and noticed that gravity and acceleration were equivalent. He didn't figure this out using math. He didn't deduce this equivalence principle from his original two assumptions about the speed of light and the laws of nature in inertial fields. He didn't even conduct experiments from the top of the tower of Pisa. He noticed it. Maybe the most amazing thing about this is that nobody had noticed it before.

After noticing the equivalence of acceleration and gravity, Einstein thought about the man in the box, and introduced the curvature idea, with the beam of light following a curved path in front of the man. The curvature of this path in an accelerated field would have to correspond exactly to the curvature of light in an equivalent gravitational field.

Einstein didn't say that he knew exactly how large objects like planets created these gravitational fields. Einstein didn't even say that he had proof that these fields existed. Rather, he assumed that they existed because of what was known about other fields involving magnetism, electricity, and light; and also, because the idea of action at a distance was contrary to his common sense. Action at a distance would be like someone trying to pull up on the box in space, from a distance, without the rope. And mainly, of course, Einstein didn't say that gravity is not a force.

The reason we sometimes hear that gravity is not a force, I think, is that we don't have enough words to accurately describe all the elements involved in whatever it is that is going on, when objects are attracted to each other. Consequently the word 'gravity' is used selectively, and arbitrarily. Sometimes it is used to refer to the gravitational fields in space alone, leaving the impression that there is nothing more to gravity than curved space. For people who had thought of space as being empty, this created a problem. Something must have happened to space, to change it from being an empty vacuum, to being something which could be curved.

Luckily Einstein had a few more things to say about space. And since the popcorn bowl is empty now, we can just use it as a cheap introduction to the next question. When we bring the empty bowl to the kitchen for a refill, have we brought the bowl's empty space along with it?

Chapter 16

SPACE

On June 9, 1952 Einstein wrote a note to be included in the next printing of his little book, 'Relativity', explaining why he had added an appendix. Here is a small portion:

> "I wished to show that space-time is not necessarily something to which one can ascribe a separate existence, independently of the actual objects of physical reality. Physical objects are not in space, but these objects are spatially extended. In this way the concept "empty space" loses its meaning."[58]

In my copy this is printed right after the preface, and is entitled: "Note to the Fifteenth Edition". The fifth appendix to which this note refers is printed at the end of the book. It is, as Einstein promised, about the nature of space, and how it is never empty. So Einstein both opened and closed this late edition of his book by telling us that there is no such thing as empty space.

---

58 Albert Einstein: Relativity, p vii.

In the fifth appendix Einstein takes us through a thought experiment, in which we have an 'empty' closed box, and we move it about. We are asked to think about the space in the box. When we move this box, does the space inside the box move with it? Does the box enclose the same space, or different space, as it moves about? What happens to the space outside the box when the box moves? Does it get into the box? What happens to the space inside the box when the thickness of the walls of the box is reduced to zero? It gets even more interesting when we put the empty box inside a larger empty box. Do the two boxes now both 'contain' the space inside the smaller box? What if we move the little box around inside the larger one? I have embellished it a bit, but hopefully not changed the question.

In giving his answer, Einstein uses the small letter 's' to denote the space inside the smaller box, and the capital letter 'S' for the space inside the larger box:

> "One is then inclined to think that s encloses always the same space, but a variable part of the space S. It then becomes necessary to apportion to each box its particular space, not thought of as bounded, and to assume that these two spaces are in motion with respect to each other."[59]

---

59 Albert Einstein: Relativity, p 158-159.

This is a straightforward answer. The spaces move. He then goes on to describe the old way of thinking, where space was empty, and space and time existed independently of each other and also independently of physical objects, as 'pre-scientific thought'.[60] Throughout history, our experiences of space and time had always been subjective, despite attempts by science to objectify them. It is a bit surprising that the idea which accomplished this objectification, bringing time and space out of the realm of subjectivity and into the realm of science, would be called 'relativity theory', since we often interpret 'relative' to mean 'subjective'. In fact, as we know, this terminology was not of Einstein's choosing. He would have preferred the term 'invariance' to 'relativity'. 'Invariance' would have been closer to 'objective'. Einstein continues:

> "The whole content of the special theory of relativity is included in the postulate: The Laws of Nature are invariant with respect to the Lorentz transformations."[61]

In other words, do the math and you will see that the rules never change. This has a very different ring to it than the phrase: "Everything is relative." Where did that come from, anyway? If we go back to our inertial frames moving about, we remember that

---

60 Albert Einstein: Relativity, p 162.
61 Albert Einstein: Relativity, p 169.

even though things in one frame looked different when viewed from another frame, we could always do the math (apply the Lorentz transformations) and figure out what the measurements would look like to an observer inside any particular inertial frame. In this way we are using mathematics and science to maintain invariance and objectivity. And we call it relativity. Go figure.

So how do our empty boxes fit with the objectification of the concepts of space, time and matter? As we saw in the above quotation, Einstein concluded that the spaces are moving relative to each other. The spaces move because the boxes move. The boxes don't move in space, since they are not 'in space'; they are 'spatially extended'. You can't move an object without moving its space. Yes, this is idea is definitely a 'modification'. Yet it makes sense due to the electromagnetic structure of matter as described by Lorentz, and the electromagnetic fields of Maxwell. Instead of discrete objects in empty space, we have electromagnetic objects within electromagnetic fields, and no discreet demarcation between the two. With matter giving off and absorbing energy, and matter and energy being equal, there are no more sharp edges between matter and space.

Space has become spaces, and the spaces are fields which move about relative to each other with varying degrees of acceleration, and we find ourselves back on familiar ground, even though our language struggles to keep up. We know how to deal with moving inertial frames, and we understand that accelerated frames and gravitational frames are equivalent and also manageable, at least by those who have learned how to do curved geometry. The only thing still missing, as usual, is exactly how an object produces a gravitational field without actual acceleration.

So it appears that what had changed in Einstein's thinking between 1916 and 1952 was, at the very least, that there was no longer any real empty space. This is not to say that we can't still maintain the idea, the concept, of empty space; it's just that, as with the unicorn, we're not likely to find any. Let's wind this up with some more of what Einstein said in 1952:

> "But it must now be remembered that there is an infinite number of spaces, which are in motion with respect to each other."[62]

So we're still talking about relative motion and things over there looking different when we look at them from over here. But now, instead of having things move in space, we have spaces moving too. And not just moving, but accelerating. And not just going in the same direction at different speeds, but going in any direction. But in order to go from the special to the general theory, Einstein had to make Minkowski's space-time diagrams adapt to these accelerating spaces and gravitational fields, and still make them fit on graphs drawn on two-dimensional paper.

---

62 Albert Einstein: Relativity, p 159.

The explanation of this which I have heard is that there had to be an imaginary line, or axis, coming up out of the paper. Transposing these new diagrams into mathematical statements would require a slight change to the Lorentz formula. Einstein needed a number to represent the imaginary axis coming up from the paper. The poor little, already misunderstood, 't' was about to be replaced with an imaginary number.

Chapter 17

SPACETIME

On many occasions I have run out of space, or out of time, but I don't remember ever running out of space-time. I wonder what that would be like. And if space and time are different parts of space-time, then if we are a little short of one, could we just replace it with a little more of the other? What proportion of space-time is space, and what proportion is time, or does it matter? My mother, who was a vegetarian, used to say, just for fun: "If we had some eggs we could have ham and eggs, if we had some ham." She never actually cooked ham and eggs as far as I can remember, but if she had done so, it wouldn't have mattered what the proportions were. As long as there had been some of each, it would still have qualified as ham and eggs. Now perhaps, even if there had been a negative amount of one of them, we could still have had ham and eggs, at least mathematically. For example, we could have each had one slice of ham, and negative two eggs. Look, Mom: ham and eggs. Mathematically, that is.

As early as 1908, before general relativity had been developed, Minkowski had already introduced the idea of space-time as a consequence of special relativity:

"Henceforth space by itself, and time by itself, are doomed to fade away into mere shadows, and only a kind of union of the two will preserve an independent reality"[63]

This comment was made to an audience of mathematicians, who would need to be able to show space and time measurements using numbers and graphs. On another occasion, Minkowski gave some common sense justification for combining measurements of space and time:

"Nobody has ever noticed a place except at a time, or a time except at a place."[64]

The word 'noticed' seems important here. Notice that he didn't use the word 'imagined'. We are talking about real measurement here. And as we know from our moving frames, if we happen to notice two clocks in two different moving spaces, the clock readings meeting our eye from these two spaces will be different from each other. Different spaces, different times. A different time for each moving space. All we have to remember now is that by 'time' we mean the clock readings meeting our eye.

63Herman Minkowski, Cologne, 1908, as quoted by Michael White and John Gribbin: Einstein, A Life In Science, p 95.
64 Hermann Minkowski, as quoted by David Toomey: The New Time Travelers, p 51.

Nothing new or magical has been added to our world, and nothing has changed. We still have invariance and objectivity.

And as far as space-time being four-dimensional is concerned, this isn't new either. We have always had three dimensions of space and one of time. With special relativity, which only involved inertial frames moving in the same direction, the four-dimensional grid was Euclidian, or flat, and the Lorentz transformation was the lens through which one looked in order to maintain invariance between inertial frames. In other words, it took into account how things looked different from different perspectives, due to the constant but less-than-infinite speed of light. And all of this was possible using a normal graph on a flat sheet of paper.

In order to go from special to general relativity, in other words, in order to apply the Lorentz transformations for real gravitational space, a little more math would be needed. The lines on the familiar Euclidian graph were no longer straight. They curved about; they rotated. Since it took a few years for Einstein to perfect the math to his satisfaction, even with plenty of help, I can only dream of one day understanding it. How do you describe curved lines to somebody, using only numbers? But not to worry. Einstein reassures us:

> "Minkowski's work is doubtless difficult of access
> to anyone inexperienced in mathematics, but since
> it is not necessary to have a very exact grasp of

this work in order to understand the special or the general theory of relativity, I shall leave it here at present,...."[65]

In numerous references, Einstein gives credit for the space-time idea to Minkowski. From a chapter he entitles "Minkowski's Four-Dimensional Space", Einstein explains the relationship between Minkowski's four-dimensional space-time and Euclid's three-dimensional geometrical space, and introduces the new replacement for 't':

> "In order to give due prominence to this relationship, however, we must replace the usual time-coordinate t by an imaginary magnitude..."[66]

And there it is, the 'imaginary magnitude' that we have been anticipating. Einstein goes on to complete this sentence, not with words, but by inserting the symbol for the square root of negative one; the imaginary number. By the time Einstein wrote these words, he had clearly accepted the idea of joining space and time using mathematics. His acceptance of the idea had not come immediately, however, as physicist Lisa Randall reminds us:

---

65 Albert Einstein: Relativity, p 63-64.
66 Albert Einstein: Relativity. p 63.

"The spacetime fabric is a very important notion. It concisely characterizes the geometry that corresponds to the gravitational field produced by a particular distribution of energy and matter. But Einstein initially disliked the idea, which had seemed to him like an overly fancy way to reformulate the physics that he had already explained."[67]

It seems that Einstein thought the space-time idea was too fancy for special relativity, but not for the general theory. Theoretical physicist Harald Fritzsch quotes Einstein on the subject of progressing from special to general relativity. These are Einstein's words:

"To put it into a more general framework, I was much helped by the form it was given by the mathematician (Herman) Minkowski, who was the first to point out the formal equivalence of the space and time co-ordinates, and who used this equivalence in constructing our theory."[68]

Without understanding how the math works, we can, nevertheless, understand what it does. Using imaginary numbers

---

67 Lisa Randall: Warped Passages, p 107.
68 Harald Fritzsch: The Curvature of Spacetime, p 1.

allows a graph on a flat piece if paper to show three dimensions of space and one dimension of time, and to depict points in space-time not only in a single direction as in special relativity, but in all directions. Just knowing what we do about the common sense behind special relativity, with things looking a little different when viewed from one inertial frame into another which is in relative motion to it, we non-mathematicians can relax knowing that this is what the mathematics in general relativity is talking about also. You can't read somebody's clock accurately, or the length of their breakfast table, unless you know not only their motion relative to you the observer, but also the shape and strength of the relevant gravitational fields.

The special theory grows into the general theory when we apply the same common sense ideas we used with inertial frames, to accelerating and gravitational frames. We still have 'invariance', after using a mathematical lens. By pointing out the 'equivalence' of the space and time coordinates, Minkowski has written into the math the facts that clock readings and length measurements depend on perspective, and that there is no preferred perspective. The mathematics has allowed the theory to progress from inertial frames to gravitational, or accelerated, frames. We understand this from our old familiar principle of relativity, which took away home plate, and which now takes away any preferred space. By using imaginary numbers, the mathematicians can compare not only inertial spaces moving in the same direction at constant speed, but all spaces moving or accelerating in any direction.

And so this new mathematical tool allows us to look in all directions, and to give equal treatment to all spaces. We can figure out what any frame will look like from any other frame. This is what this tool is intended to do; to describe the invariance of the laws of nature in every imaginable moving frame. The measurement of real places-at-times, and times-at-places, and their mathematical relationships with each other, is all that this tool is intended to do, when used properly.

But tools can also be used improperly, and when they are they can be dangerous. We saw earlier what could happen when someone divided by zero. The more powerful the tool, the more dangerous they are. What we need to remember, in order to safely use this tool, this formula with the imaginary number, is that just because a certain thing is imaginable, this doesn't mean that it really exists. Using mathematics we can imagine the concept of a negative amount of something, such as eggs. This doesn't mean that there must be a real kitchen somewhere, where some real mother's son is eating a negative number of real eggs, and by doing so making himself even more hungry.

The more powerful the mathematics becomes, and the less we simple folk understand it, the more vigilant we must be in insisting on being shown what its purpose is, and where it actually attaches to reality. Safety first, as they say. It's not so bad when we talk about going back and forth between spaces, since it is understood that we are always going forward in time. But what happens when we apply our imagination, and our imaginary numbers, to time itself?

Chapter 18

TIME TRAVEL

Until I was seven or eight years old, my family lived with Aunt Ruby, who had a sandbox. With apologies to the patient reader, since it isn't actually pertinent to our theme here, I must put off explaining the importance of the sandbox for a moment, and mention that Aunt Ruby was a saint.

Everybody treated Aunt Ruby as if she were some sort of invalid, just because her hands were twisted up kind of funny, and she had big swellings under her thick stockings where her ankle bones should have been, and maybe a few other such frivolous reasons, like not being able to stand up straight. I never really noticed any of these things when I was a kid. To me she was the tree of life, balanced high on a stool in the kitchen in her heavy black shoes and shapeless blue house dress, straining scalding hot preserves through a huge cheesecloth bag into a galvanized metal bucket on the floor. I never knew how she had managed to hang that bag from the ceiling, but I did know where the resulting little jars of delicious jam were kept, in neat rows on shelves in the cold cellar. I was often the lucky one who got to run down and get another jar when we ran out, or when a visitor came to the house.

At Christmastime she would ask us kids if we wouldn't mind helping her prepare gift packages for some other kids in some far away country. This involved counting assorted candies and other treats into hand folded red cellophane pouches, each with its own ribbon and bow. There were hundreds of them. The only other things that I had seen people collect for kids in far away places were old shoes and hand-me-downs. But Aunt Ruby's packages were more beautiful, and prepared with more care, than anything we ourselves would unwrap on Christmas morning. It wasn't as if Aunt Ruby felt sorry for those far away kids; it was more like she just loved them and respected them so much that they deserved something extra special. Yes she was a saint.

Aunt Ruby was my grandmother's sister, so she was actually my great aunt. She had a big house and no children or husbands, and a big yard with a big garden and maybe the biggest chestnut tree in the city. Underneath the chestnut tree was a sandbox that looked like it had been there forever, which didn't strike me as unusual when I was young, even though, as I mentioned, Aunt Ruby had no children.

So we used it, because nobody else did. There were four of us, and there were four triangular boards nailed on to the corners of the box for us to sit on. Each of us would build a house and driveway, and help with the roads and gas stations and parks and lakes in the middle. During construction of the roads, if we wanted to visit each other, we would just have to pick up our toy cars and put them in someone else's driveway. There was no travel time. Time was suspended while the car was lifted from one house to the other, since it wouldn't have been right to include the car's flight in the sandbox activity. We never thought of ourselves as time travelers, yet we could move from one frame of reference to another faster than the speed of light, and it might be time for breakfast even though we had just had lunch. When things are imaginary there are no restrictions on taking instantaneous trips in space or time.

Since then I have somehow become prejudiced against the whole idea of time travel, which is probably why I have been reminiscing about Aunt Ruby instead of getting down to business and talking about it. Outside of the sandbox, the idea of time travel just sounds silly. And yet I want to give the topic serious treatment, and at least try to understand why serious people even talk about it. Maybe if I could read this chapter before writing it, I would see how badly it turned out, and decide not to write it at all. It would save me a lot of time. And it would keep me safe from the danger of being disrespectful or flippant, if it isn't already too late.

But not only do I not know how to take a trip forward in time, and read the unwritten chapter; I wouldn't even know how to plan for it. Should I take a suitcase? Will I need stronger reading glasses by the time I get there? How much time will it take me to get from the old time to the new time? And which kind of time will I be using up during my trip; past time, or present time? Will

I have to do anything dangerous, like go faster than the speed of light, or wear a cape? Will I need to shower when I arrive? Can I take pictures along the way? Are we there yet?

But now I have done what I did not want to do with this topic; be flip. Much more fitting to be respectful and serious here; there are smarter people than me who know more math than I do, and who have thought about this a lot more than I have, and who think that some sort of time travel may be possible. Okay, then, being serious, what legitimate questions would I ask about my proposed trip through time. Let's see, should I take a suitcase?

It's no good. I can't help myself. No matter how smart the time travel advocates are, if they are allowed to say that I can travel into the past, or into the future, then I get to ask my questions. I want answers. And mostly I want answers when they mention Einstein. As far as I can tell, Einstein didn't say that time travel was possible, either as a result of the special, or the general theory.

I think that the most common reason people think that special relativity implies time travel is due to the fact that they misinterpret the original thought experiment about light beams and moving frames of reference. They seem to forget that in this experiment everybody and everything was going in the same direction, and at constant speeds, and that the whole point was that the laws of nature do not change just because the distance between a clock and a person watching the clock increases.

As long as this distance increases steadily, the time between the ticks of a clock appears steady. The distance must keep increasing steadily to keep the ticking slow. The faster the distance increases between the clock and the person reading it, the slower the ticks are seen by the person. It's common sense. Light speed is limited. But as we have seen, this doesn't mean that time has slowed down anywhere. And the person watching the moving clock can't jump instantaneously up beside it by moving faster than light, and grab the clock while it still looks slow, and then claim to have traveled back in time.

My apologies here again; all of this moving and ticking and traveling of light between clock and person is hard to visualize while reading about it; it is easier to picture while driving or doing the dishes. It's just that we need to be aware of the danger of imagining that we can jump between frames, like kids in a sandbox.

When general relativity is used as the basis for time-travel theories, we often see the same kind of thing happening, in my opinion. People want to jump instantaneously from one gravitational frame to another. And indeed we know that the clocks will read differently in a different gravitational field. Since gravity can slow down the motion of a clock, clocks in different gravitational fields will tick at different rates. Does that mean that anything that affects the rate of ticking of a clock affects the flow of time? In that case why not just immerse a big mechanical clock in a bathtub full of molasses? We could even work out a mathematical formula for the slowing of the gooey clock, as proof. Then we could jump into the bathtub with the clock and travel back in time. It seems silly. But surely there's more to it, since lots of smart people take the idea of time travel seriously.

One person who does take time travel seriously, and who is more qualified to have an opinion about it than I am, is Ronald L. Mallett, PhD. I read his book "Time Traveler" with gratitude. My gratitude was for his honesty and sincerity, and the way he does not talk down to the reader. He just tells his story. His text, written with Bruce Henderson, is most definitely 'poignant and powerful', as Kristine Larsen, professor of physics and astronomy, says on the back cover.

When Mallett was about eleven years old his father died of a sudden heart attack. The course of young Ronald's life, after this sad event, was set when he came across a book about time travel. He resolved to build a time machine and go back in time and save his father's life. Here he describes what he read, at the age of twelve, about Einstein and the Lorentz transformation:

> "Lorentz said time, represented by the letter t in his equation, could be affected by motion."...

And:

> "Furthermore, Einstein's special theory of relativity, utilizing the Lorentz transformation, said that time slows down the faster you move. In

other words, time slows down for a moving clock......To my eager twelve year old mind, Einstein's theory of special relativity served as an inspiration because it seemed to suggest that time travel was possible, which I took to mean that a time machine was also a possibility. "[69]

Mallett went on to become a physics professor, teaching among other things, Einstein's relativity theory. He never gave up his quest to build a time machine. Does he still believe that 'time slows down for a moving clock'? The answer to this seems unclear. In a footnote on Lorentz, Mallett doesn't seem to be saying this:

"In 1904 he developed his Lorentz transformations, mathematical formulas that relate space and time measurements of one observer to those of a second observer moving relative to the first."[70]

In this footnote Mallett gives a much more conventional explanation of how the Lorentz contraction formula was applied to observations of space and time. The formula's purpose is to 'relate space and time measurements'. Exactly so. Invariance. The meaning of this footnote, it seems to me, is very different from

69 Ronald L. Mallett: Time Traveler, p 23 & 24.
70 Ronald L. Mallett: Time Traveler, p 200, #4.

saying that time slows down for moving clocks. I can't help wondering how different Mallett's life would have been, if at the age of twelve, his information on Lorentz had been received as he himself describes it in this footnote. Or, if at that early age, he had also been told about the idea that all motion is relative, and that the 't' for 'time' put into the formula by Einstein referred only to clock readings in one frame as perceived from another frame, by somebody who had to wait for the clock signal to arrive. Young Mallett might have just thought that it was interesting that things over there look different when viewed from over here, instead of thinking that Lorentz and Einstein were saying that someone could build a time machine.

Yet as of the time of writing his book in 2006, he was apparently still working on building a time machine, and still saying that it was based on Einstein's relativity theory. Here is how he described what he once said to a reporter:

> "As I recall, I was being compared to Doc Brown, the mad professor in Back To The Future who invented the Flux Capacitor to travel back in time. In exasperation, I finally said, "Look, I am not a nut. This is not Ron Mallett's theory of matter. It's Einstein's theory of relativity."[71]

---

71 Ronald L. Mallett: Time Traveler, p 161.

The time machine he was working on at the time of writing his book was based upon something called 'frame dragging', which, if I understand correctly, involves laser light traveling in a circular path in one direction, creating and 'dragging' gravitational fields, meaning space-time, along with it. He concludes that time travel into the past may be possible, but only back to the time when the time machine is first turned on. This means that he won't be able to go back and save his father after all, unless he finds a time machine built before or during his father's lifetime.

Mallett's idea seems, to me at least, to be based upon the notion that space and time are physically joined, as components of space-time. If matter creates a gravitational field, and energy is matter, then energy in the form of light also creates a gravitational field. So in this way light can bend, or drag, not only space, but time along with it.

But as I understand things, Minkowski joined space and time mathematically, not physically, and the equations were used only to transpose and therefore 'equate' measurements taken in different reference frames, thus showing the invariance of the laws of nature. Using the right formula, one could calculate how much a certain thing would weigh, or how much a certain light beam would bend, in various gravitational fields at a certain time. Everything is always explainable using common sense, and the laws of nature always hold. Measuring a certain space at a certain time does not physically join the two, any more than taking a photograph would. Added to this is the fact that the math works for imaginary situations just as well as for real situations.

Or should I say 'imaginable'. It is interesting to note that in the spinning time-machine, when the space-time is dragged around, it is only the time, and not the space, that goes into negative numbers. From where I sit, one reason for this may be that our minds are able to imagine negative time, but not negative space. But that's just from where I sit.

What I do like about Mallett's idea is that it doesn't seem to involve instantaneous sandbox-style frame-jumping. The object or person for whom time will slow down is inside the contraption before it is turned on. But a major jump has still been made, in my opinion, and that jump is away from Einstein. For the thing to work, the version of Einstein and Lorentz' work being invoked is the version that Mallett learned when he was a boy. For the thing to work, motion has to slow down time. And according to the way I read Mallett's own footnote, this is not what Lorentz or Einstein said.

So if Mallett's time machine ever does work, I personally think that he should take the credit for it himself, and not tell reporters that it was Einstein's idea. And if by any chance it works well enough to go back to a time before it was built, I would like to borrow it from him. I miss my Aunt Ruby. If, on the other hand, Mallet's machine doesn't achieve time travel at all, there are still some other ideas we could try.

Chapter 19

WORMHOLES

"The wormhole can be used as a time machine but you've got to get through before it scrunches you up."

These are the words of physicist and prolific writer Dr. Paul Davies, who was delivering a public lecture in the year 2000, at the Royal Society in England. I found this lecture online, while looking for stuff on wormholes and time travel.[72] About a year after giving this lecture, Dr Davies wrote a book called "How To Build A Time Machine".[73] The reason you have to be careful not to get 'scrunched up', according to Dr. Davies, is because the wormhole opens and closes, presumably at regular intervals. He mentions the movie 'Contact', based on a novel written by Carl Sagan. In the movie the hero uses a wormhole to take a shortcut through space, and must go through it at just the right moment.

Davies says that the wormhole can be used not only as a shortcut through space, but as a time machine, since time on either

---

72 http://vega.org.uk/video/programme/61.
73 published by Penguin/Viking, 2001.

end of the wormhole can be made to flow in opposite directions. This was the part of his talk that most interested me. I admit to holding my breath for the part in which he was saying that time slows down for traveling clocks and twins, although I realized that he was explaining Einstein in a way that seems to work for some listeners, who can't give up on the idea of a stationary home base. Davies also helpfully distinguishes between two types of theoretical time machines. There are those which reverse the flow of time, as we saw with the frame-dragging machine earlier, and those which simply take you from the present and deposit you into the past, without undoing the time or the activities in between. The wormhole based time machine is of the second type.

The idea seems to be that a wormhole is a black hole with two openings opposite each other, or two black holes connected by a sort of tunnel. Time can go in different directions at the two openings. This isn't something that anyone has seen with a telescope or measured with a reversible clock; rather it is something that some people have seen in the equations for general relativity. Apparently the equations still work whether time goes forward or backward. Whichever way time goes, the equations are still correct. They are still correct because the standard of correctness in mathematics is self-consistency. Hold the equations up to a mirror and they still work.

In 'real life', the standard of correctness is more demanding. The image one sees of oneself in the mirror is self-consistent, but it is not correct. If I stand in front of the mirror and touch my right ear, the person looking back is touching his left ear. He is also much older than me, by the way. Maybe he has traveled forward in time after all. Just kidding. Anyway, if we recorded all the measurements for the live person, and for the reflection of the person, the measurements would all be correct, but only one set of measurements would describe something real.

So we need a better connection between mathematics and reality. More mathematics isn't enough. We are reminded of what Einstein said about geometry; it can be self-consistent without corresponding to reality. Here Einstein distinguished yet again between correct geometry and truth:

> "The concept 'true' does not tally with the assertions of pure geometry, because by the word 'true' we are eventually in the habit of designating always the correspondence with a 'real' object..."[74]

Not that there's anything wrong with using mathematics to look for real objects or conditions; the black hole itself is a perfect example of a successful search which originated in the math. But in cases where we don't find something real, something physical, it's just more mathematics. The danger is always that somebody will prematurely objectify the math, and assume the existence of its

---

74 Albert Einstein: Relativity, p 4.

physical counterpart, without any physical evidence. The 'object' being described in the mathematics of the wormhole is space-time. We are looking at both ends of the wormhole at the same time, and saying that they are different parts of a physical thing, and that time behaves differently at opposite ends. It is as if we are saying, from outside the sandbox, that 'right now', time is going in one direction on this side of the sandbox, and in the other direction on the other side. All we have to do, in order to go back in time, is to jump from here to there, right now.

But one can't assume that because Einstein used the mathematics of space-time, he was saying that 'right now' different spaces have different times. On the contrary, here is what Einstein did say:

> "The four-dimensional continuum is now no longer resolvable objectively into sections, all of which contain simultaneous events; 'now' loses for the spatially extended world its objective meaning. It is because of this that space and time must be regarded as a four-dimensional continuum that is objectively unresolvable ... "[75]

To me this means that you can't think of the math as describing a single, static physical structure. You can't stand on the

---

[75] Albert Einstein: Relativity, p 170.

'spatially extended' grass beside the sandbox right 'now', and divide the sandbox 'objectively into sections'. This brings us back to the common sense of relativity theory; Einstein uses the mathematical idea of space-time to explain the invariance of the laws of nature. It always comes back to the idea that if motion is relative, there is no preferred reference frame, including no view from outside of the sandbox where you can judge that time is going this way over here and that way over there. All you can ever say is that from this part of the sandbox, things look different in that part of the sandbox. The whole purpose of the curved geometry of space-time is not to show that various space-times are different; it is just the opposite. The mathematics explains away the perceived differences and shows that all reference frames follow the same rules. Just as with the inertial frames of reference in relative motion, we can transpose the information from one gravitational frame into the terms which suit another frame. Electronic piano keyboards have 'transpose' buttons which allow the player to raise or lower the pitch by half-tones. You see one note being played, but hear another. When you know about the button, everything makes sense.

Physicist Lee Smolin addresses the difficulty of handling time and space using mathematics. He describes the earliest known graphing of space and time together as 'the scene of the crime', showing that it goes back at least to Descartes and Galileo:

"Around the beginning of the seventeenth century, Descartes and Galileo both made a most wonderful discovery. You could draw a graph, with one axis being space and the other being time....In this way, time is represented as if it were another dimension of space....This 'spatialization' of time is useful but may be challenged as representing a static and unchanging world....We have to find a way to unfreeze time-- to represent time without turning it into space."[76]

By calling this the 'scene of the crime', I take Smolin to mean that while it may be ok to join space and time mathematically, it is not okay to then go ahead and treat the result as a single, static objective reality. If one assumes a single objective reality, then yes, one will get past, present and future existing simultaneously, like a stack of photographs taken at different times in the same place. Drive a nail through the stack of photographs, and the nail represents time. If this is what objective reality is like, then yes, one might find a way to travel up and down the nail, backward and forward in time.

The analogy that is sometimes used here is traveling across borders from one country or state into another. When you cross a border, the place you left behind is still there, still just as real as the

---

76 Lee Smolin: The Trouble With Physics, p 256-257.

place you are entering. Some would say that time travel is like this; the past still exists after we leave it, and the future is already there waiting for us to step into the picture. Presumably, if we are to find Julius Caesar alive and well, this will be the way we will find him. Unfortunately for him, however, the photographs of his demise are also already in the stack.

On the other hand, if one assumes a continuously evolving present, with infinite spaces bobbing about like small boats on an ocean, one can use the mathematics to orient oneself. There is still no objective, correct time, from which various clocks deviate. Events which are seen as simultaneous in one space may be seen as happening at different times when viewed from another space, due to the differing amounts of time it takes light to travel from the event to someone watching. Neither can be said to be the correct observation. None of the bobbing boats can claim ownership of the 'right' time, or even of the 'right' sequence of events. The best that we can do is to figure out how and when someone in another space might see something; if we want to, we can use our transpose button or formula to know what someone else is seeing.

Using the math to transpose the physical data, we understand why our observations differ, but we never reach an objective resolution and choose a solid, static version of space-time. Before we can drive a nail into the stack of photographs, we have to stack them up. The question is, if everyone's photos look different, how do we choose which stack to drive the nail into?

The most common demonstration of the wormhole effect involves a sheet of paper with an ink dot on either end. The shortest way to get from one dot to the other is to fold the paper in half so that the dots come close together. One is asked to imagine that the two-dimensional sheet of paper represents our three-dimensional universe. Bend the paper so that the dots are close, join the dots with a wormhole, and then take the resulting shortcut to the farthest reaches of the universe. Part of the sales pitch for the folded paper idea involves an ant crawling on the piece of paper. The ant is oblivious to the curvature of the paper, and thinks that the only route between the two dots is along the surface of the paper. Since the ant is oblivious to the curvature in the paper, the reasoning goes, then we humans might be oblivious to a curvature of space, or an extra dimension. To me this seems like a pretty weak sales pitch. I think that I am more aware, and hopefully also more imaginative, than an ant; but I still can't imagine extra dimensions, or even three dimensions with a shortcut which would by-pass one of them.

Of course we do represent three dimensions on two dimensional paper all the time, with photographs and drawings, and for some people, with mathematics. But this doesn't mean that we can fold the universe in half and connect the stars. And even if we could bend space around far enough to bring two distant places close together, then the two places would be, well, close together. We wouldn't need a wormhole at all if we could bend three dimensional space in half like a piece of paper. We would just be able to travel a shorter distance through space. Yes, the extra gravity may scrunch us up, but we shouldn't be allowed to say we were going extra distance, even if these words were to be our last.

One physicist who is well known as an expert on black holes and wormholes, and who was the one the folks who made the film 'Contact' consulted for advice, is physicist Kip Thorne. Here is a bit of what he said about black holes, for which there is physical evidence, and about wormholes, for which there is none:

> "Whereas black holes are an inevitable consequence of stellar evolution (massive, slowly spinning stars, of just the sort that astronomers see in profusion in our galaxy, will implode to form black holes when they die), there is no analogous, natural way for a wormhole to be created. In fact, there is no reason at all to think that our Universe contains today any singularities of the sort that give birth to wormholes...; and even if such singularities did exist, it is hard to understand how two of them could find each other in the vast reaches of hyperspace, so as to create a wormhole..."[77]

Once scientists do find something physical to sink their

---

77 Kip S. Thorne: Black Holes & Time Warps, p 486.

teeth into, it is amazing how much they find out about it. Here is just a tiny taste of the knowledge acquired so far about the formation of black holes, as described by Frank Close:

> "In heavier stars, once the nuclear fuel has been used up, there is no heat pressure to resist gravity and the dense iron core collapses catastrophically; the outer layers fall in on the core and bounce off, sending gigantic shockwaves outwards ... .
> Left behind is a compact neutron star or a black hole;...."[78]

This is real stuff. Get out your binoculars and your calculator and roll up your sleeves. But when it comes to wormholes serving as time machines, we're not at that point by any means. And as for time machines in general, as Stephen Hawking purportedly said, if it is in our future to build time machines, then why haven't we come back and visited ourselves by now? As far as we know now, time machines exist only as imaginative speculation. Not that there's anything wrong with imaginative speculation.

---

78 Frank Close: The New Cosmic Onion, p 41.

Chapter 20

IMAGINATIVE SPECULATION

Things can be logically connected in at least two different ways. Something can be logically consistent with something else, in the sense that the two things are both independently true and they don't contradict each other. This is very different from saying that things are connected by logical necessity, meaning that if one of them is true then the other one must be true also.

The philosopher Spinoza liked to write long sequences of statements which followed each other by logical necessity, each new idea being deduced from the preceding one. It was as if all of the ideas had been folded up inside the original, and all one had to do was to follow the rules of logic, in order to unravel all of the hidden, necessary implications of the original idea. If the first statement was true, and the rules of logic were not broken, then all the rest of the statements had to be true also.

An undergraduate course on Spinoza which I attended was particularly tedious and boring until someone found a statement which was logically consistent with the others, but not logically necessary. After that the participation level increased noticeably;

139

the class had turned into a kind of treasure hunt, as we looked for more examples of consistency without necessity.

We have seen, or we have taken on faith, that there is a lot of mathematics which is consistent with relativity theory, and which has inspired lots of imaginative speculation. We have seen that there are imaginary time machines that take us from one place in time to another, without being able to change what has happened. Other imaginary time machines would allow us to change what has happened, and Superman to save Lois Lane. Some imaginary time machines would only take us back to the time when we first turned them on, like Ronald Mallet's project. There are yet other imaginary time machines which would enable us to go back in time, but any changes we made would result in the universe splitting, so that we would only change the outcome in one universe, while the other one progressed to produce ourselves.

Another imaginative and speculative idea is that past, present and future exist simultaneously, but that we can't experience them all at once, or choose which part to experience next. Rather than using up time, and rather than living in some sort of river of time, we move along a pre-existing, fixed roadway of time which connects past present and future, the way a thruway connects cities. The trouble is that we can't turn around and we can't get off the thruway. So, according to this theory, while Julius Caesar is alive and well right now in the past, unfortunately we are unable to get there from here, or even to communicate with him.

Besides theories of time travel and simultaneous time frames, there is serious and wonderful imaginative speculation about multiple universes and multiple dimensions, all claiming links to Einstein. Similarly, ideas about stringy or loopy particles, gravity waves, gravity particles, dark matter, dark energy and gravity which can leak between universes, or between unseen dimensions, all claim similar roots. Years of study have gone into something called 'landscape' theories. Combined with something called the 'anthropic principle', I think these suggest that every conceivable universe exists, and the only reason that we live in this one is that this is the only one that we could live in. This is reminiscent of the children's song: "We're here because we're here". I'm sure I am leaving out lots of important speculative theories, and probably misrepresenting the ones I haven't left out. The point is that there are lots of them. And that is a good thing, to paraphrase another popular expert, Martha Stewart.

The more the merrier. As long as everybody is clear that it is speculation; sort of a giant brainstorming session. Brainstorming sessions are not only fun, but very useful and fruitful in very many areas of human endeavor. Even when Einstein's ideas are used as a springboard, most of the time nobody would object. After all, the whole idea of brainstorming is to have one idea inspire new ones, with only the act of inspiration itself serving as a sufficient link between them.

When it comes to brainstorming based upon Einstein's fundamental ideas, it could be that the only people who object to it are the ones like me, who doggedly insist on first understanding what Einstein said, all by himself, before hearing about all the exciting new ideas which he inspired in others. It's not that we're not interested in the other stuff; we just want to know the difference between what Einstein said, in the first place, and what is now known, really, and what is now known, maybe.

On a recent visit to yet another physics exhibit, I was excited to hear a young presenter proclaim that dark matter has definitely been found. The exhibit was in a busy outdoor town square, with lots of curious onlookers asking questions, and it was difficult to get close enough to the booth to even read the posters. It is amazing how many people are interested in this stuff. But above the noise of the crowd, several times I did hear this young voice announcing the great discovery. Well this was good news indeed, and I went home and wrote it down.

When I heard in this way about the the discovery of dark matter, I incorrectly assumed that the discovery had been made in space, with some sort of probe.. But sometime within the next week or so I realized that the 'discovery' was found only on the mathematics. Of course I did not question the mathematics, yet I was not satisfied with the mathematics alone, as proof of something like this. My next trip to our public library netted a fairly new and very intriguing book by physicist John W. Moffat, in which I found, among many other helpful insights, these two statements:

> "Indeed, since August 2006, there has been a groundswell of opinion ....that the case is almost closed, that the existence of dark matter has been proved."[79]
>
> (but)
>
> "One should not draw premature conclusions about the existence of dark matter without careful study of alternative gravity theories and their predictions..."[80]

It turned out that there was logical consistency in the dark matter hypothesis, but not logical necessity. There were other possible explanations for the 'missing matter' in the universe. Moffat went on to explain that merely showing that a theory can explain something, like the theory of dark matter explains the rate of expansion of the universe, isn't enough. It may be logically consistent with everything else going on, and the mathematics may work, but if it lacks both logical necessity and physical evidence, then it is just another good idea. It is imaginative speculation.

---

79 John W. Moffatt: Reinventing Gravity, p. 166.
80 John W. Moffatt: Reinventing Gravity, p. 168.

Einstein also came up with an idea to explain the rate of expansion of the universe. The mathematics showed that the universe should have a different amount of mass that it has, in order to hold itself together. So Einstein just inserted the necessary number into the math, and called it the 'cosmological constant'. He was embarrassed by this afterward, and called it a mistake. This was not because it was ever disproved, but rather because it could not be proved. It was logically consistent with what was known about the universe, but he had just made it up. The 'dark matter' idea which is gaining so much acceptance today sounds to me very much like Einstein's cosmological constant. It is consistent with physical observations, and the math works, but it is not logically necessary and there is no physical evidence for it yet. Of course, this doesn't make it necessarily wrong, either; it just makes it another good idea.

In my mind I still have the image of that young student in the booth at the outdoor science fair, explaining dark matter to the crowd of onlookers. Or is 'explaining' the right word? Given the setting, it was almost as if he were selling something. Kind of refreshing, actually; good for him, we need more of that. But, as I play back the scene in my mind I am tempted to distort it a bit, and imagine him saying to the curious onlookers: "It slices, dices, chops and juices; and it's guaranteed not to chip, crack, peel or break for a full year." Yes, he was a good salesman, and there's nothing wrong with that. In fact it's a rare quality in a physicist.

Unfortunately, in this case the product proved to be somewhat less than advertized. Within a week after taking it home, it did show a couple of chips and cracks. Without physical evidence or logical necessity, and with only logical consistency, it was just more imaginative speculation. Not that there's ... well, you know.

Einstein Didn't Say That

Chapter 21

UNCERTAINTY

Most everybody has heard of the uncertainty principle. 'Heisenberg's uncertainty principle', that is. And most people are pretty sure they know what it means, if you ask them about it in a casual sort of way. Some people are more comfortable with the idea than others. The odd thing seems to be that the people who are most certain that they know what the uncertainty principle is, are the ones who have heard the least about it.

Some people have only heard that you can't measure the speed and position of a particle at the same time. Or is it momentum and direction? Or position and momentum? Whatever. Most people are satisfied to know that you can't know everything. Others have heard only that the act of measurement interferes with the thing being measured. Still others, that a particle has neither a particular position nor a particular amount of momentum until it is measured. Or even that nothing exists until it is observed. Physicist John Wheeler apparently went even further, saying that:

"Without an observer, there are no laws of physics."[81]

_____

81 John Wheeler, quoted by Ronald Mallet: The Time Traveler, p 127.

As long as one has heard only one of these wordings, or is satisfied that in some vague way they all mean the same thing, one's certainty about uncertainty remains intact. The rest of us, like Goldilocks, are left to find a version that is palatable to us; not too hot and not too cold. Maybe we will mix some of the choices together, or maybe we will even invent our own version. Or we may decide that we should just forget about the uncertainty principle, because Einstein didn't like it. But as soon as we do that, someone will say that it was Einstein who interjected uncertainty into physics in the first place, with his work on Brownian motion.

Robert Brown was the botanist who in 1827 first observed pollen grains, which he had collected from flowers, jiggling about under his microscope. What was making them jiggle? After ruling out any unsteadiness in his equipment, he thought that the movement might have been due to the fact that the pollen grains were living things, and called the jiggly bits 'animalcules'. But when he put non-living matter under his microscope he still saw the same action. Everything jiggled. While people of his day were interested in seeing this phenomenon, nobody really knew what it meant. As time passed, there was some speculation that the motion may have been due to the movement of the atoms inside the particles. The idea was then linked to the mechanical theory of heat, in which the amount of heat depended on the intensity of the motion of the atoms in a given object. Ludwig Boltzman introduced statistics into the mix, saying that if hot and cold portions of a gas were mixed together, the random collisions among the atoms would average out the temperature. He came up with some mathematics which described this. It was all about statistics, randomness, and the probability of atoms within particles colliding with each other. But he stopped short of developing a mathematical formula which would make predictions about the actual movement of the whole particles, from place to place, due to the jiggling. It was Einstein's 1905 paper on Brownian motion which accomplished this, giving in the process, wide acceptance to the very idea of the existence of atoms.. Einstein's paper involved the probability of how far particles would move in a given time, due to the action of the atoms inside the particles. There is that word, 'probability'.

It was Einstein's idea also, put forth in another paper written the same year, that the smallest particles were nothing more than packets or 'quanta' of energy. This was, as we have seen, an expansion upon Lorentz' idea of the electromagnetic structure of matter. So it seems that in 1905 we have Einstein already talking about particles in terms of probability, and matter in terms of quanta. Sounds pretty inexact. And 'inexactness' was the word that Heisenberg had originally preferred for his uncertainty principle.

But when Heisenberg's idea landed on the scene, Einstein was not impressed. He wrote to a colleague:

"Heisenberg has laid a large quantum egg."[82]

_____

82 David Lindley: Uncertainty, p 116.

And also, in perhaps one of his most often quoted statements, he said that God didn't play dice.

Astrophysicist David Lindley, in his thorough and meticulous book "Uncertainty", walks us through the whole process of the origin, the development, the controversy, and what I would call the marketing, of the uncertainty principle. This was one instance in the development of physics where a lot of thought was put into marketing. Heisenberg's idea was carefully packaged and marketed, less by Heisenberg himself than by others. Some would say that his original idea had to be altered, in order for it to fit into the packaging. But none of the marketing worked on Einstein. David Lindley describes how Heisenberg at one point tried to appease Einstein by comparing the situation to Einstein's own introduction of relativity theory, in which events could be interpreted differently by different observers. Here Lindley describes Einstein's reaction:

> "Einstein wasn't buying it ... because special relativity accounts for the discrepancies ... An underlying objectivity persists."[83]

Yes this has a familiar ring to it. Invariance. The laws of nature apply equally in all frames of reference.

---

83 David Lindley: Uncertainty, p 132.

As far as I have seen, no one ever did sell Einstein on Heisenberg's uncertainty principle. But as far as I can tell, it wasn't the uncertainty itself that Einstein objected to; what Einstein objected to was the reason for the uncertainty. When Einstein had been calculating probabilities related to Brownian motion, the single reason that there were only probabilities, instead of certainties, was that there was no equipment which could measure the motion of the atoms inside the particles.

Heisenberg's reason for the uncertainty, on the other hand, was not the lack of adequate measuring tools. Physicist Janna Levin sums up the difference succinctly:

> "For instance, the uncertainty principle is not due to an experimental lack of precision but instead results from a cardinal ambiguity. The particle does not have a definite location or a velocity but rather is in a superposition of possible states of being."[84]

We have to remember that the particle we are talking about is by definition the smallest thing that qualifies as something material. It is a packet of energy, a quanta, at the point where it becomes matter. It seems that we are trying to describe in physical terms something which does not quite yet have physical properties.

---

84 Janna Levin: How The Universe Got Its Spots, p 61.

Talk about your fundamental problems. This brings us back full circle, to where we started on our small journey. Einstein was trying to reconcile what was known about the nature of light with the principle of relativity, and bring these two seemingly incompatible ideas together, into the realm of common sense. In the process he uncovered some astonishing truths about the nature of matter and gravity. At the same time, our language was inadequate to even clearly describe the new theory to each other, much less to elaborate without confusion on its implications.

The emergence of the uncertainty principle was similar in that it presented a conundrum which was not explainable in common sense language. The difference is that the very claim of the uncertainty principle seems to be that, at some level, things are not explainable. The fundamental property of the most fundamental part of existence is that it doesn't have fundamental properties. How is our poor language supposed to handle this situation? David Lindley reports that Heisenberg said to his superior, Neils Bohr: "Our words don't fit."[85]

The question of whether light consisted of waves or particles had been going on for a long time, and was still going on, in Heisenberg's day. Heisenberg had apparently preferred to think of matter in terms of particles, rather than waves. His problem was that, by the time a particle was observed to be in a certain position, there was no single explanation for how it got there.

---

85 Heisenberg as quoted by David Lindley: Uncertainty, p 150.

You see Uncle Henry from the old country at the family reunion, but you don't know which route he took to get here, or how long it took him. But you will ask him later, and he will remember, hopefully. If he doesn't remember his trip, it will just be due to his failing memory. But if we were to apply the uncertainty principle, in its various forms, to Uncle Henry, then there would be other possible reasons for him not remembering his trip. Maybe this information is not knowable. Or maybe his plans weren't made until the instant he arrived. Or maybe there was no trip. Or maybe there were many trips existing together in a 'superposition of possible states'. Maybe all we can ever know is that he probably just came to the reunion because he always does.

To me this situation compares to the one Einstein found himself in, when trying to reconcile seemingly incompatible ideas in 1905. There had to be a way to explain things in common sense terms. The uncertainty principle introduced a similar situation, which has yet to be resolved in a comparable, common sense way. We are still at the stage of speculation. It was inevitable that one day we would eventually find the smallest bit of whatever it is that stuff is made of, whether we call it matter or energy, wave or particle. When we do get down that far, if we're not there already, how are we supposed to explain how that little bit got there?

Don't ask me. But the best explanation can't be that things are unexplainable. So I'll just wait for the next genius who insists, like Einstein did, on things making sense.

Chapter 22

RELATIVITY

Suppose I said that if I win a fifty million dollar lottery this winter, then I will buy a fifty foot white sailboat next summer. Suppose also that, being unreasonably optimistic, I go looking at sailboats and asking lots of questions about the size of their sails and their speed. Now imagine that, because of all this questioning, I have a flash of insight about the relationship between the size of the sail, the wind speed, and the amount of energy converted to motion of the boat in the lake. Imagine also that this information is received as a very important breakthrough, and is expressed in a simple formula that everyone recognizes: E = SW (energy equals sail area times wind speed).

You can be sure that before the lottery takes place there will be much talk about this important discovery and what it means. "He said that the formula proves he is going to win." The formula has been tested experimentally; he will win for sure." "He said that if he wins ten million dollars he will buy a ten foot sailboat." "According to him, bigger sails make the wind stronger." "He said that fifty foot sailboats are white." "He said that the wind curves the sails, and the curvature pushes the boat."

Now add to this the fact that I am going to assume the validity of Galileo's 'principle of relativity', which states that the laws of nature are the same in all inertial fields. If this assumption is true, then it is clear that my formula will work for any sailboat on any lake in any part of the universe, even though the speed of various planets through the universe will vary. Now, if we want to, we could add the fact that boats which are far away from us will look small, and will also look slow. We have all seen sailboats in the distance; they look like little toys, almost sitting still.

Because I invoked Galileo's 'principle of relativity', people will say that all of these things are parts of my new theory of relativity. "He said that the equation proves that far away boats go slow." "He said that the wind speed is constant because everything else is relative." Then somebody will spend the weekend before the lottery translating all of this into math, calculating what it all means, or what it might mean. "According to the formula, wind speeds of less than zero will push the boat backwards." And what is the relationship between moving inertial frames? Is one boat 'really' smaller? Why are there different opinions about this? What does the color of the boat have to do with the date of the lottery? It must be in the math somewhere. After all, sometimes we do discover new truths hidden in the math, and there's nothing wrong with that. But what will you say to someone who politely asks: "Excuse me, what did he say?", or, "Excuse me, what didn't he say?"?

Quite a mess, isn't it? Well, it looks to me like that's pretty much how Einstein's 'relativity theory' has been packaged. It has inspired lots of ideas, some good and some not so good, and it seems as if all the ideas have been tossed into a big cardboard bin like we would find in a discount store, and we all take turns taking things out, examining them, and putting them back, looking for that single piece which will explain all the rest. Each of us may choose a different piece, and try to explain all the others in terms of that one. It gets even more confusing when some of the pieces look very different, but have the same label.

Here is gravity; it is curvature. It is not a force. Here is gravity again; it is a force emanating from a large mass. And this piece of gravity is a wave; or no, a particle I think. This one is also called gravity; it is a field. There is an extra sticker on it: "Caution! Contains force".   Here is the 'principle of relativity'.   It means things are the same. Here it is again; it means things are different. What to do? Well, in my own case, my own 'happy hours of contemplation' have resulted in the realization that there were a lot of things in that bin that Einstein didn't say. A lot of it was nice stuff, but it was just in the wrong bin.

Einstein didn't claim to have proven Galileo's 'principle of relativity'. It was an assumption. Einstein didn't claim to have proven that the speed of light is constant in all inertial frames. This was also an assumption. Maxwell's equations for the speed of light apply only to light in a vacuum in a single frame. If there were no moving frames, and no 'principle of relativity', E would still be equal to MC². Similarly, if there were only one lake and one sailboat in the universe, my equation would still work. We could realize either truth, without any need to call it relativity theory.

Then why did I call my sailboat thing relativity theory? I didn't. They did. And furthermore, if I had chosen a name for it, I wouldn't have chosen that one. I would more likely have chosen 'invariance theory'. Maybe if Galileo had called his idea 'the principle of invariance', there might have been a little less confusion. Words are the packages that we put ideas into, and if they don't quite fit, the ideas get distorted. It can be like trying to put a kitchen appliance back in its box, along with all the attachments, and all that styrofoam, without having to remove the handles.

Einstein said that he began by trying to reconcile two ideas, when he discovered that E=MC². At the end of the previous chapter on uncertainty, I intimated that he had succeeded in reconciling those ideas. This statement of mine, upon further reflection, turns out to be a bit of a 'stretcher', as Huckleberry Finn would say. It was indeed during his attempt to reconcile the relativity of motion with the constancy of the speed of light, even for a single ray of light as measured from fields moving at different speeds, that Einstein noticed that the electromagnetic structure of matter meant that it contained a measurable amount of energy.

158

Yes, he was thinking about fields in motion, when he noticed that $E=MC^2$. But did Einstein really explain, or claim to explain, what it was, physically, about the nature of light, which made it possible for the same ray of light to propagate through two differently moving frames at the same speed relative to both? I don't think so. Einstein didn't say: "The speed of light is the same in all frames because time slows down for the other guy.". Einstein did say, on the other hand: "IF the speed of light is less than infinite, and constant, in all inertial frames, and IF the laws of nature are the same in all inertial frames, THEN clock readings will depend upon relative motion. He described how relative motion would affect the waiting time for clock signals.

Similarly, Einstein's formulation of the 'equivalence principle' resulted, not from thinking about Galileo's relativity idea, but rather from leaning back too far in his chair. Once again, he had simply noticed something very profound. Acceleration and gravity were equivalent. From this realization he got into lots of mathematics involving curvature, and in this way joined this new general theory to the original special theory. If measurements of moving objects come up short, then, theoretically, if you spin a bicycle wheel, its circumference measurement will shrink a tiny bit. What happens to the spokes of the wheel if the wheel gets smaller? Well, they become curved of course. This makes for a nice mental picture, showing the link between the special and general theories, for those of us who can't see it in the math.

Einstein didn't say that he had proven that his two original assumptions were correct. Einstein didn't say he knew why acceleration and gravity were equivalent, or how the force of gravity was transmitted, or how light could always propagate away from you at the same same speed, even if you chased it faster and faster on your bicycle. But in the process of contemplating on these things he surely did notice some profound unquestionable truths. Physicists are still looking for ways to test Einstein's original two assumptions. Many physicists think that the expansion of space can make light exceed its own speed limit, as in the big bang. Many physicists are still looking for a wave or particle which may carry the force of gravity. But physicists are not questioning the validity of $E=MC^2$, or the equivalence of gravity and acceleration, as far as I can tell. Who knows, maybe there will be a single new discovery which will prove Einstein's assumptions, and explain the uncertainty thing.

But now my imagination wants to take over. I'd better quit in case I'm ahead. I just wanted to mention some of the things that Einstein didn't say, and identify some of the questions which still need answers. Having put everything into separate bins, we may experience a momentary peaceful feeling, like the one we get when we finally finish cleaning the basement. Once we have found some solid ground, we can safely let our imaginations run free, while we sit back, relax, and just reflect on things.

BIBLIOGRAPHY

Arianrhod, Robyn: Einstein's Heroes, Oxford, New York, 2005

Barnett, Jo Ellen: Time's Pendulum. Plenum Press, New York, 1998.

Bodanis, David: E=MC². Doubleday Canada, 2000.

Bodanis, David: Electric Universe. Crown Publishers, New York, 2005.

Bova, Ben: The Story Of Light. Sourcebooks, Naperville, Illinois, 2001.

Brennan, J. H. : Time Travel A New Perspective. Llewellyn Publications, St. Paul, MN. 1997.

Brian, Denis: Einstein: A Life. New York, John Wiley & Sons, 1996.

Chaisson, Eric: Relatively Speaking. W.W. Norton & Company, New York, 1988.

Close, Frank: The New Cosmic Onion. Taylor & Francis, New York, 2007.

Cropper, William H: Great Physicists. Oxford University Press, New York, 2001.

Deutsch, David: The Fabric of Reality. Penguin Press, New York, 1997.

Einstein, Albert: Relativity, The Special and General Theory.

Wings Books, New York, 1961.

Einstein, Albert: Out Of My Later Years. Wings Books, New York, 1956.

Falk, Dan: In Search Of Time. Thomas Dunne Books, New York, 2008.

Folsing, Albrecht: Albert Einstein: A Biography. England, Penguin Books,1997.

Fritzsch, Harald: An Equation That Changed The World, University of Chicago Press, Chicago, 1994.

Galison, Peter: Einstein's Clocks, Poincaré's Maps. W.W. Norton & Co., New York, 2003.

Genz, Henning: nothingness: the science of empty space. Perseus Books, Reading, Mass., 1994.

Gleick, James: Isaac Newton. Pantheon Books, New York, 2003, Random House, Toronto, 2003.

Gleick, James: Faster: The Acceleration of Just About Everything. Pantheon Books, New York, 1999.

Gribbin, John: The Origins of the Future. Yale University Press, New Haven and London, 2006.

Gribbin, John: Schrodinger's Kittens and the Search for Reality. Little, Brown, & Company, New York, 1995.

Guillen, Michael: Five Equations That Changed The World, Hyperion, New York, 1995.

Hawking, Stephen: A Stubbornly Persistent Illusion. Running Press, Philadelphia, 2007.

Hawking, Stephen: Black Holes and Baby Universes. Bantam Books, New York,1993.

Hawking, Stephen: On The Shoulders Of Giants. Running Press, Philadelphia, London, 2002.

Hawking, Stephen: The Universe In A Nutshell. Bantam Books, New York, 2001.

Henshaw, John M. Does Measurement Measure Up? The Johns Hopkins University Press, Baltimore, 2006.

Herbert, Nick: Faster Than Light; Superluminal Loopholes in Physics. Nal Penguin Inc, New York, 1988.

Hirshfeld, Alan: The Electric Life of Michael Faraday. Walter & Company, New York 2006.

Isaacson, Walter: Einstein His Life and Universe. Simon & Schuster, New York, 2007.

Jones, Roger S.: Physics for the Rest of Us. Contemporary Books, Chicago, 1992.

Kane, Gordon: The Particle Garden. Helix Books, Reading, Mass. 1995.

Kaku, Michio: Einstein's Cosmos. New York, W.W. Norton & Company, 2004.

Kaku, Michio: Physics Of The Impossible. Doubleday, New York, 2008.

Kirkland, Kyle: Particles and the Universe. Facts On File, Inc, New York, 2007.

Levin, Janna: How The Universe Got Its Spots. Weidenfeld & Nicolson Ltd., London, 2002.

Lindley, David: Uncertainty. New York: Doubleday, 2007.

Lloyd, Seth: Programming the Universe: Random House, New York, 2006.

Magueijo, João: Faster Than The Speed Of Light. Perseus Publishing, Cambridge, Ma., 2003.

Mallet, Ronald L.: Time Traveler. Thunder's Mouth Press, New York, 2006.

Mermin, N. David: It's About Time. Princeton, Princeton University Press, 2005.

Moffat, John W. Reinventing Gravity. Harper Collins, New York, 2008.

Newton, Roger G: From Clockwork to Crapshoot, a history of physics. The Belknap Press of Harvard University Press, Cambridge, London, 2007.

Oerter, Robert: The Theory of Almost Everything. Pi Press, New York.

Pais, Abraham: Einstein Lived Here. Oxford University Press, New York, 1994.

Pais, Abraham: Subtle is the Lord...New York, Oxford University Press, 1982.

Parker, Barry: Einstein's Brainchild, Prometheus Books, New York, 2000.

Perkowitz, Sidney: empire of light. Henry Holt & Co., New York, 1996.

Randall, Lisa: Warped Passages; Harper Collins New York, 2005.

Segrè, Emilio: From Falling Bodies To Radio Waves. W. H. Freeman & Co., New York, 1984.

Seife, Charles: Zero: the biography of a dangerous idea. Viking Press, New York, 2000.

Smolin, Lee: The Trouble With Physics. Houghton Mifflin Co., Boston, 2006.

Struble, Mitch: The Web Of Space-Time. The Westminster Press, Philadelphia, 1973.

Susskind, David: The Black Hole War. Little, Brown and Company, New York, 2008.

Thorne, Kip S.: Black Holes & Time Warps, W. W. Norton & Company, New York & London, 1994.

Toomey, David: The New Time Travelers. W.W. Norton & Company, Inc. New York, 2007.

Von Baeyer, Hans Christian: Maxwell's Demon. Random House, New York, 1998.

Waltar, Alan E.: Radiation and Modern Life. Prometheus Books, Amherst, N.Y., 2004.

White, Michael & Gribbin, John: Einstein A Life in Science. Penguin, New York, 1994.

Wilczek, Frank: The Lightness Of Being. Basic Books, New York, 2008.

www.ingramcontent.com/pod-product-compliance
Lightning Source LLC
Chambersburg PA
CBHW060028210326
41520CB00009B/1049